EVERYDAY EXTRAORDINARY

EVERYDAY EXTRAORDINARY

A SCIENTIST PONDERS A LIFETIME OF MAGICAL, BIZARRE, AND PARANORMAL EXPERIENCES

BARRY MARKOVSKY

Prometheus Books
Essex, Connecticut

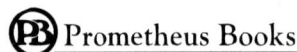 Prometheus Books

An imprint of The Globe Pequot Publishing Group, Inc.
64 South Main St.
Essex, CT 06426
www.globepequot.com

Copyright © 2026 by Barry Markovsky

All rights reserved. No part of this book may be reproduced in any form or by any electronic or mechanical means, including information storage and retrieval systems, without written permission from the publisher, except by a reviewer who may quote passages in a review.

British Library Cataloguing in Publication Information available

Library of Congress Cataloging-in-Publication Data available

ISBN 978-1-4930-9379-3 (cloth)
ISBN 978-1-4930-9412-7 (ebook)

♾️ The paper used in this publication meets the minimum requirements of American National Standard for Information Sciences—Permanence of Paper for Printed Library Materials, ANSI/NISO Z39.48-1992.

For Jeri

CONTENTS

Introduction: The Fourth Response ix

Chapter 1: Boy's Nose Amputated by Cousin 1
Chapter 2: Bed Head. 6
Chapter 3: Hanukkah Boys Wait Up for Santa 22
Chapter 4: You Sneeze, You Die 30
Chapter 5: U.F. . . . Oh? 43
Chapter 6: All Rise. 56
Chapter 7: Onward Christian Scientists 68
Chapter 8: He's Pulling Her Leg 81
Chapter 9: Astronomy 95
Chapter 10: You Made the Earth Move 108
Chapter 11: Dowsing the Dowser 115
Chapter 12: The Yogi Has No Robes 128
Chapter 13: The Power of the Pyramid 144
Chapter 14: ESP: Can It Be? 155
Chapter 15: Are We *Real* Doctors? 174
Chapter 16: That's the Spirit! 185
Chapter 17: Numbers Game 197
Chapter 18: Circling the Square 211

Epilogue 237
Acknowledgments 239
Notes 243
Index 269

INTRODUCTION: THE FOURTH RESPONSE

What's going on here?

I'm sitting on the couch reading. A low morning sun is streaming in through the windows behind me. From the corner of my eye, I notice my daughter quietly descending the carpeted stairs about ten feet to the right. She stops at the bottom. I raise my head and look over to see what she wants.

No one is there.

Has anything like this ever happened to you? If so, how did *you* respond?

My first reaction was completely natural: *Weird. I have no idea what that was.* I shrugged it off and tried to keep reading.

I couldn't concentrate, so my second response was, *What could that have been over there?* No obvious answer presented itself. But I couldn't let it go. I had to wonder: *Maybe a ghost? I can see how some people would think that.* Still, unsatisfying.

Third, I questioned myself. *Maybe there's something wrong with my eyes. Or maybe I'm hallucinating.* This also made no sense. I never hallucinated before, and my eyes worked just fine.

Looking back, my responses up to this point were predictable, ticking three boxes in sequence:

1. Uncertainty: *I don't know what just happened.*

2. Looking outward: *What* was *that thing over there?*

3. Looking inward: *Was it my eyes? My brain?*

INTRODUCTION: THE FOURTH RESPONSE

Getting nowhere, I moved on to a fourth response:

I wonder if I can make it happen again.

I sat in my original reading position, this time paying extra attention to my peripheral vision. Sure enough, I saw it again. And soon, it was obvious what had happened. More on that in a moment.

This book's title implies that extraordinary events like this are commonplace. But what makes them *seem* extraordinary is usually a matter of inexperience with the phenomenon, or lack of familiarity with its probable causes.

The following chapters are chronological, and each opens with an extraordinary phenomenon described in a "Frontstage" section. Each is a true story about an eerie incident, uncanny coincidence, bizarre claim, or other wondrous event. It may be something I personally experienced or observed, or that was told to me by others who experienced it directly. I'll apply some creative license to fill gaps in conversations and contexts, and I'll alter the names of some of the people still living. Otherwise, the descriptions stick closely to my recollections.

Around half of the earlier stories involve different kinds of religious claims—from Santa Claus to miracles. It's not that I have an axe to grind and need to debunk religious beliefs. Rather, it's that so many of the eerie, bizarre, uncanny, and magical claims I encountered in my younger days were connected to my own or others' religions. You'll see what I mean.

After each story's Frontstage section is a peek "backstage"—the "fourth response," if you will. It invites you to see the claim from a vantage point that pokes and prods it, considers alternative explanations, and accounts for it in down-to-earth terms. The explanations will often surprise and delight you, as they did me when I first learned about them.

My favorite writers are so interesting and entertaining that I'm largely unaware of how much I learn from them while reading. We'd all be lifelong students if only school were more like their books. I try to capture some of that spirit in these pages. There are no academic prerequisites, and later chapters don't depend on earlier ones, should you want to skip around. Yes, here and there you'll encounter bits of physics, social science, physiology, and so on. But I make a little bit of science go a long way.

If you're a seasoned skeptic, you'll notice a kinder and gentler brand of "debunking" than you've likely encountered before. You might even find the approach useful when offering your skeptical perspective to those who are skeptical of your skepticism.

If you have unshakeable beliefs in one or more of the claims we examine, I hope you'll feel both challenged and respected. I may poke some fun at a claim if there's obvious fun to be poked, but always it's to inform and entertain, never to mock or embarrass. I'm simply offering perspectives, explanations, and evidence you may not have seen before. I'm grateful you're willing to look and consider.

Over the years of reading and teaching these subjects, I've heard all the reasons people give for and against their validity. I always approached each topic with an open mind as I learned more about it. But as wiser people have recommended, my mind was not so open that my brains fell out. A claim warrants a closer look when, if true, it would be unprecedented, nonsensical, or in violation of physics. Even more so when it has important consequences for the believer's health, finances, or relationships.

With any extraordinary claim, most people fall between the extremes of true believer and died-in-the-wool skeptic. But just being moderate in your belief may not be wholly satisfying. Maybe you're skeptical about astrology or ghosts but can't back up your position when confronted by a believer. Or maybe you feel that some of the ESP or UFO cases you've heard about are plausible, but you haven't seen a rigorous analysis done by trained researchers. I've written this book for you: one who may not be squarely on the fence, but at least you can *see* the fence from where you sit.

You may assume of me, "This guy is so biased! He thinks science explains everything." The approach I take is *scientific skepticism*, and whether or not it warps judgments or closes minds is a very reasonable question to raise.

Ethically and practically, before I publish anything, I have to consider the possibility I'm deluding myself. Maybe science as a whole is self-deluded. But the chances are reduced because, unlike any other belief system ever invented, science has built-in safeguards against bias. Also, we scientists are anything but closed-minded about what's possible. *All* major theories in science at one time went against the accepted wisdom. It can take time, but when evidence accumulates against even the most firmly held beliefs, science eventually lets go of the old and ushers in the new.

On a more personal level, I'd love to be the first to demonstrate beyond a doubt that ESP works, astrology has predictive powers, or our planet is being visited by extraterrestrials. Many mainstream scientists feel the same. Our institutions reward us for out-of-the-box discoveries. I have to be open, at least a little, to the possibility that people who claim to cure diseases with "vibrations," or talk to dead ancestors, can really do so. But to move my belief meter's needle from "Unlikely" to "Likely," a claim must be supported with both a logical explanation and solid, repeatable evidence. So yes, I'm biased. I'm biased against believing illogical claims, as well as any claim that lacks good evidence.

What about the argument that scientists think they can explain everything? Honestly, *any* scientist saying such a thing would have to be a fool. Every day, science explains more than it did the day before and shows no signs of slowing down. This proves that there is always more to explain. Science moves forward; it improves, but it's never finished.

Conducting tests and interpreting results requires some professional training. So does distinguishing good research from bad. If you want to get to the bottom of an extraordinary claim, plunging into a Google rabbit hole isn't the same as doing a competent literature review. Do-it-yourself research of any kind is never advisable without appropriate skills in research methods and, very often, statistics.

It generally isn't obvious when a proponent of nonsense lacks research skills or a scientific disposition. It's easier than you might think to sound authoritative while spouting rubbish. This book won't solve all misinformation problems. I do hope it provides a few cautionary tales while fostering healthy skepticism and critical thinking skills.

In the following chapters, I'll serve as your helmsman, helping you navigate the vast, choppy oceans of astonishing claims. But before we set

INTRODUCTION: THE FOURTH RESPONSE

off, let's return to the scene of my apparition, or whatever it was. This is what happened "backstage."

When I checked to see if I could re-create the vision on the stairs, it worked perfectly. Every time I stared down at the book, I saw movement from the corner of my right eye. And every time I looked up, nothing was there. I tried it without my glasses. No more apparition. When I put the glasses back on, I slowly tilted my head this way and that and moved the glasses to different locations on my nose, changing the distance between the lenses and my eyes.

When I bought my glasses, an extra fifty dollars for the anti-reflective coating seemed like an unnecessary luxury. Maybe I should have sprung for it. That morning, the streaming sun behind me filtered through some skinny young cypress trees waving in the breeze just outside the window. The trees reflected off the inside right lens of my glasses and lined up perfectly with the image of the stairs coming through that same lens. The tree images were peripheral, distorted, out of focus, and ambiguous. Like most brains, mine dislikes ambiguity. So, it *interpreted* the reflection as someone coming down the stairs.

A last word about scientific skepticism: Some friends and students have told me that, for them, science takes all the fun out of mysteries. I can understand why they feel this way. But I believe they're in the minority.

Science has plenty of real mysteries yet to solve, but among them are extremely few popular claims of the magical, miraculous, supernatural, and paranormal. What makes these so popular is that the claims get disseminated far more widely than the explanations. When a perception is deemed *extra*-sensory, a flying object *un*-identified, or an apparition *other*-worldly, these titillating labels explain literally nothing beyond admitting that the observer doesn't know what the thing is. It's okay if they're happy believing something's a mystery when it's not. It's fine if they lack the curiosity or time to solve it. But seeing the mechanisms that give rise to such extraordinary things is far more entertaining and empowering.

Let's pull back some curtains and see what lies backstage.

CHAPTER ONE

Boy's Nose Amputated by Cousin

Those who don't believe in magic will never find it.
—Roald Dahl[1]

Frontstage
Boredom tightens its death grip. The grown-ups are shoulder to shoulder in the kitchen, doing things with food, engaged in pointless conversations. Now and then, one of them squirts out of the swinging door into the dining room and puts something onto the table we can't touch yet. My brother, three male cousins, and I sneak out of their fields of vision and begin to rumble and slam our way around the big old house like little pro wrestlers. There will be injuries and property damage. Happy Thanksgiving.

We're visiting relatives on my mother's side. Twenty-eight miles away in a Boston suburb, it's the furthest I've been from home my whole life. All three years of it.

Out of the blue, Willie, my oldest cousin, decides it would be hilarious to fool me with the I've-got-your-nose trick. He yells, "Hey Barry!" and approaches to enact his evil plan. He's nine years older than me, and I like the attention.

"Huuuuuhhh? What's this?" he asks, frowning and squinting at a point somewhere between my eyes, which cross as they follow his loosely fisted hand moving closer to my face. I feel his knuckles lightly brush the tip of my nose.

Then, *magic*.

"Look! I've got your nose!" Willie announces. He holds up his clenched fist with a fleshy thingy jutting out between his index and middle fingers. I'm horrified as he introduces, right in front of what used to be my nose, undeniable evidence of my dismemberment. God, how will I survive with a hole in the middle of my face?

Willie breaks a sly smile, now fully satisfied with the terror he's wrought. I smile back without amusement or comprehension. Then slowly, dramatically, he opens his hand to reveal how he accomplished the illusion. Before I can say anything, he's already spun around and bolted, hoping to see if the trick works as well on the family cat.

Launching one of my earliest experiments in a long research career, I use trial and error to reproduce the trick with my own hand. It only takes a few tries. But while fumbling around with my fingers, I'm struck by two simple facts: First, a thumb tip looks ludicrously un-noselike when protruding between the fingers. Second, if my cousin actually had yanked my nose off my face, it probably would have hurt. Somehow neither consideration entered my mind at the moment of truth.

Willie fooled me completely despite the obvious clues that it was only a trick. But I don't feel duped for long. Bigger kids are supposed to make littler kids feel like chumps whenever opportunities arise. I'm just happy to learn the new trick. I toddle off to see if I can find and fool Miss Floofenstein before Willie gets to her.

Backstage

Even if you knew and trusted me, what assurance would you have that my memories of these events are the least bit accurate? There is this thing called *hyperthymesia*,[2] an ability to remember practically everything one has ever experienced. It's rare, and I'd remember if I had it, which I don't. What I do have, like most normal people, is a good memory for significant experiences. But for even the best of memories, our brains automatically do some gap-filling to smooth over the content and images we recall. I'm pretty sure this thing with my nose really happened, though the part about the cat may have been embellished or totally concocted. I'm not 100 percent sure they even had one.

When my cousin cut off my nose to spite my face, it must have been emotionally impactful for that memory to have burned into my brain so permanently. Emotions make experiences memorable.[3] Most people have few recollections from toddlerhood. In my case, being whipsawed between horror and glee was enough to lock in my memory of the faux amputation.

Emotion is a double-edged sword. Just as it contributes to the preservation of cherished experiences, it can warp and skew perceptions and then lock in false memories stemming directly from those misperceptions. Emotions felt long ago can make today's memories of nonevents seem absolutely real.[4]

Children's brains are the focal point for many developmental psychologists. Their research suggests it's virtually impossible that I'm accurately recollecting events from so early in life. The experience of retrieving memories *feels* like replaying a video recording, but that's not at all how it works. What's worse, just as actual video recordings can be misleading by offering only one perspective on an event, human memory is even more fallible.[5] While recording with our senses, we may miss crucial information or divide our attention. Once a memory is stored, we may lose key bits of information to stress, aging, or injury. When replayed, the memory may be tainted by cross-feed from other memories or interference from aging, chemicals, and distraction.[6]

These factors wedge themselves between memories and events, and we're blissfully unaware of them. Nevertheless, whether a memory aligns with reality or is tainted by distortion, we generally assume it's accurate.[7] To see this, we need to look no further than the couple disagreeing over who said what in a recent argument, or siblings stunned to discover they have radically different memories of the same profound family events from childhood.

Beyond the quirks of memory, another factor deserves some attention. The mind of a small child goes through rapid changes over relatively short periods. A two-year-old might respond to the nose trick with a shrug of the shoulders and think, "Okay. Bye-bye nose. Can I run around and play some more?" Most two-year-olds can identify their own eyes, ears, mouths, and noses but may understand nothing of their location's permanence or the implications of their loss. Only months prior, they'd have failed Mr. Potato Head.

One time, my dad pretended to pop out one of his eyeballs and clean it in his mouth. He concealed the empty socket by closing that eyelid after the extraction. Rolling the eye around in his mouth, you could see the bulge in his cheeks as he moved it side to side. Finally, he spat the eye into his hand, popped it back into the socket from whence it came, and fluttered open the lid to reveal a shiny, clean orb. Such a trick would have been pointless until I reached a certain level of cognitive development. As it was, I was old enough not to be fooled but still young enough to be amused. Mere months later, it was eye-rollingly dumb, just another groaner in an endless stream of dad jokes. (What I wouldn't give now for one more dad joke straight from the source.)

Until the age of two or three, children's cognitive lives float freely on the currents of a magical sea. People and objects pop into and out of existence. Television is a window through which they watch small people, animals, and sentient puppets live their lives. I remember believing our cat to be more emotionally intelligent than my brother or parents. And unlike any of the humans, Fuzzy cared deeply about me. Cartoons were real. Magic was real. And because the grown-ups told us so, angels, devils, gods, and heaven were real, too.

Why wouldn't they be?

Around this tender age, I asked my mother where daylight comes from. In a moment of distraction, impatience, or mischief, she explained: "Sunlight comes from the downspouts on the sides of houses." It makes absolutely no sense, but I believed it absolutely. I even recall feeling a status boost when relaying this advanced meteorological knowledge to my playmates, who accepted it just as unquestioningly.

I clung to that belief longer than I care to admit. One day, the evidence of my ignorance must have overwhelmed my smug self-assuredness, and I put away the childish notion of downspout daylight. Plenty of other ridiculous beliefs remained fully intact.

I generally trust my memory of the nose trick because it's consistent with how a typical three-year-old would experience the event. But for the sake of telling this story, I have to accept that my replayed memory is only a reconstruction from strung-together fragments and connections between images, sounds, relationships, objects, people, language, contexts, expectations, and other experiences. Some of it may be true, but some may have been filled in by my brain during recording, storage, or replay. I'll never know the whole truth.

To get a sense of this reconstructive quality, take a moment and snap a mental picture of your own Thanksgiving table, whether last year's or from decades ago. Hold onto that picture for a moment.

Now think about this: Where is the camera located with which you took your mental picture? Most people place it somewhere behind and above them and can see themselves within the frame. Did you? The interesting thing about this is that if you're in the picture, you couldn't have possibly created your memory from that perspective. It must be a reconstruction of remembered elements, with yourself inserted for good measure. What's disconcerting about this is how easy it is to "remember" an image we never saw, viewed from a perspective we never had.

I say all these things about memory because they're applicable to every chapter that follows. The glitches of our memories unwittingly affect every recollection that pops into our heads. I try to keep this in mind. It's a point worth remembering.

CHAPTER TWO

Bed Head

Dreaming permits each and every one of us to be quietly and safely insane every night of our lives.
          —William Dement[1]

Frontstage
Dreams. They can be fun, scary, and even supernatural. What do they mean, if anything?

Scene One. In the darkness of half-sleep, I see a little girl walking outside in the snow and howling wind. She's alone, crying, shivering, and lost. My perspective shifts, and her face appears. It's my five-year-old daughter. I shudder and pull the comforter tight to my chin. I could be imagining more pleasant things. Why do I fixate on such heart-tugging scenarios? I know she's safe in bed, but I get up and check anyway.

Tiptoeing, I pause at her slightly ajar door and peer through the crack. She's asleep on her side, the covers moving up and down with her breaths. She'd be invisible if it weren't for a sliver of moonlight sneaking around the side of her window blinds, painting a stripe from her shoulder to her foot. I sigh, top off my heart, and shuffle toward the kitchen on autopilot.

Scene Two. Falling back asleep, a wisp of a memory begins to unfold. I'm a little boy, maybe eight, waking up one summer morning. Early light enters from the two windows in my brother's and my bedroom. With no reason to leave my cocoon, I loll around in a drowsy haze. I plot a strategy to get Mom to make me a hot breakfast, which probably won't happen.

It's more likely I'll lumber into the kitchen and find a bowl of Frosted Flakes already on the table, waiting for its splash of cold milk. But it's okay. I'm with Tony the Tiger. They're *gr-r-reat*!

Lying on my back, head slightly propped on a pillow and chin against my chest, I can make out the closed door to the kitchen over my blanketed toes. There's a wooden chair to the left of the doorway. Last night I draped my shirt over it when I changed into PJs. With the light now starting to come up, I see the entire shirt, a bit out of focus, slouched on the chair. Suddenly, the shirt spasms and bolts upright.

It has my attention.

I stare wide-eyed, except for blinking every few seconds, because my morning eyes are crunchy and sensitive. Will it move again? While waiting to find out, I scan the room and take in its details. I pinch my thigh and tug my hair to see if I'm truly awake. I focus once more on the shirt and think: *Maybe I can make it move again. Is this even possible?*

I focus on the left sleeve hanging by the side of the chair. Almost immediately, the cuff rises up to shoulder level as if to wave, the sleeve bending naturally at the elbow. I let the left arm drop down and turn my attention to the right. I mentally urge the right sleeve to do what the left sleeve just did. It obeys. Now in full command, I play with the shirt a little longer before bringing it to a grand finale: a convulsive wave radiating up the front, ending with both arms flapping. Finally, it relaxes into its original position on the chair.

I'm just a kid, and it's just a shirt, but I think I found my superpower.

Scene Three. Here's another one I'll never forget. It's a few years before the convulsing shirt. I wake up during the night. I'm lying on my left side, looking into the room. My brother is sleeping in his bed across the way. Something wispy gradually materializes a few feet from my face. It's a grayish-white cloud the size of a serving platter. Its edges churn slowly in small, turbulent curls.

Before I can grasp the fulsome weirdness of what I'm seeing, an object gradually comes into focus in the middle of the cloud. It's the face—no, the full three-dimensional head—of an elderly man. I wonder if it's a dream but think it's probably real because I'm aware of asking myself if it might not be.

The head looks at me and starts talking. I prop myself up on my left elbow and strain to understand. None of it makes sense. He may as well be speaking a foreign language. I'm not frightened or confused, only frustrated. He's talking right to me, but I can't tell what the old man-head is saying.

Within a few minutes, the apparition fades and disappears. I go back to sleep.

Come morning, the memory of the experience is crystal clear, but I speak of it to no one. It's my first ghost.

Could it have been a grandfather? Mine were both old men when they died, and I never got to meet them. They immigrated in the early 1900s: one Russian, and one Lithuanian. The only other old people I know are their widows, my grandmothers, whom I see smoking lots of cigarettes at family gatherings. They lunge at me like cobras, smother me with hugs, grab my face with both hands, and pinch my cheeks with gnarly vice-grip fingers. They croak Yiddish and broken English at me, but I never know what they say unless my parents translate.

The floating man-head's garbled speech sounded much like my grandmothers', so it makes perfect sense that one of their dead husbands contacted me from beyond the grave. What other explanation could there be?

Maybe next time he shows up, I'll have figured out a way to communicate with him. We can exchange stories about our worlds—his flight from persecution and poverty in the old country, and my obsessions with peppermint stick ice cream, spaghetti and meatballs, and Captain Kangaroo.

Scene Four. I'm running from a monster as fast as my little legs can carry me. But it's closing in fast. It snorts and growls as its footfalls grow louder. I'm terrified. When it's close enough that I feel its hot breath on the back of my neck, I suddenly realize: *This is only a dream. It can't hurt me.* So I try something radical: I stop abruptly and spin around to confront the beast. It skids to a halt, and we face each other. The thing towers over me, but I look up, and our eyes lock. With feigned confidence, I begin a conversation.

"Hi monster! I'm Barry. What's your name?"

Shocked, it recoils. Then, miraculously, its demeanor changes. All signs of menace and aggression melt away. The danger is diffused. I've made a new friend.

Scene Five. My little girl sometimes wanders around the house after nightmares she can't remember. Usually, she arrives at our bedside and stands there, torn over whether to wake us up. But she doesn't have to, as her sobbing rouses our empathy.

I grumble. Sleepiness suppresses my paternal compassion for the moment. But her mom's response is immediate and kind: She whispers soothing words even before her eyes open. Under our daughter's gaze, she lifts her head, sits up, reaches for a hug, then gets up to escort our daughter back to bed. Along their way, I can hear the start of a sweet little conversation. The last thing I hear is the child's giggle, probably as her mother tickles her while tucking the sheet and blanket around her small body.

Scene Six. As a child, I had the same nightmare several times. I'm standing in the kitchen of the family apartment. All looks normal except for one thing: There's a huge wash basin brimming with mashed butternut squash where the gas range should be. My mom sometimes prepares a batch of this bright orange "food" for dinner as a departure from the usual mashed potatoes. I hate the stuff, but I'm forbidden to leave the table until I eat a glob the size of a large ice cream scoop.

I know what's about to happen. I make a beeline for the back door, but before I cross the threshold, the sink explodes. The blast knocks me to the floor, and all goes dark. I slowly open my eyes, still dreaming, but everything's different now. It's like when Dorothy opened her door to the Land of Oz. But this is no Technicolor landscape. The apartment's interior, the world outside the windows, and myself are covered in thick layers of orange muck.

I wake up in my darkened, monochrome room feeling lucky to be alive and squash-free.

Scene Seven. Of course, I have other childhood nightmares, but one in particular stands out as profoundly disturbing. I bolt upright in bed, covers falling to my lap, screaming and crying inconsolably. Mom and Dad instantly appear and sit with me. They ask, "What happened? Did you have a bad dream?" Through convulsive sobs, I answer, "I don't know. I don't *know!*" They ask again, and I give the same answer. It's like a brain surgeon removed the memory of the nightmare, leaving behind only the horror and panic. The one saving grace is that it never happens again.

BACKSTAGE

In dreams, we can star in magical adventures with insane storylines, unfettered by special effects budgets. But unlike the movies, we never know whether our night's journey will deliver comedy, drama, romance, action, or horror. At least the ticket price is fair.

The scenes above are loosely ordered by degree of control and consciousness—most to least. Their common themes are sleep-related experiences, the mind, and childhood. Here we'll go Backstage to better understand how the scenes played out.

Scene One. You've probably experienced this: You're in bed but can't sleep. Stressful thoughts enter your mind seemingly on their own, leaking into the voids you'd rather fill with calm. Your child is in danger. You've been treated unfairly at work. You watched a war documentary and can't get the images out of your head. Some people can turn it off. Others use white noise generators, drugs, bedtime story apps, or meditation. But there are no panaceas, especially for parents.

A few years before I dove into fatherhood, a friend told me, "Having kids has brought out the absolute best in me—and the absolute worst." He explained how he'd not expected to plumb new depths of love. Nor had he expected his children to unveil new dimensions of fear, anxiety, and frustration.

When we care deeply for someone, it's bad enough to involuntarily imagine them in pain or peril. But when it's our own child, love and parental bonds make the thought of their suffering unbearable. Awake in bed with the day's distractions subsided, it's no wonder parents sometimes tumble into such fearful thoughts.

Stress and trauma are usually the most cited reasons for insomnia. Other causes include medical issues and distractions in the sleep environment. As a result, more than one in five Americans have problems getting enough sleep.[2] The health consequences can be serious. In the short term, the diminished capacities that come with sleep deprivation are familiar to all of us. Being awake for twenty-four hours is equivalent to having an illegal blood alcohol level of 0.10 in its effect on coordination and judgment.[3] Potential long-term effects are frightening: hypertension, diabetes, obesity, depression, and cardiovascular disorders.[4]

We'll lie awake because we worry about money, deadlines, love, or family. Then we worry about not sleeping or about worrying too much. Fortunately for most of us, there's no harm in an occasional restless night beyond annoying a bed partner or nodding off in a meeting the next day.

I didn't sleep consistently through the night for much of my childhood. My parents, camp counselors, and friends' parents at sleepovers always told me I *should* sleep through the night. Often, I didn't. I'd lie awake, stressed because everyone else was asleep. *What's* wrong *with me?* was a frequent theme in my insomnolent thoughts.

The burden of others' expectations doesn't end with childhood. Sleep pundits say we need eight or more hours a night—a number so embedded in our culture that people feel anxious when they come up short. But even if true as an average, eight hours is more than many of us need. I get five or six per night and rarely feel deprived. And aside from what may be evident in these pages, I don't think it's causing any mental deficits.

Scene Two. My dancing shirt hallucination happened after I'd just been fully asleep and while still on my way to waking up. This is a good opportunity to discuss what happens in our heads in the hours between hitting the hay and greeting the day.

Most of the time we're asleep, there are no dreams. On a typical night, we pass through a series of stages that clever experts named "1, 2, 3, 4," and "REM." The last is named not for the 1980s alt-rock band but instead for the rapid eye movements (REM) that accompany it. Our focus is on REM sleep.[5]

Pre-REM stages progress from lighter to deeper slumber. But they're just the warmup act. REM is the incubator where most dreams hatch, despite accounting for only around 20 percent of sleep time. Yet, the reasons for dreaming and REM sleep are somewhat murky and complicated.[6] *Something* is happening neurologically. What or why isn't so clear. Evidence points to several possibilities, including memory filtering and consolidation, learning, and removing toxins from the brain.[7] But *dream* deprivation research shows mixed findings insofar as potential harm.[8, 9]

Much more is known about what the body does while the mind dreams. The most dramatic phenomenon is sleep paralysis.[10] While the sleeper's eyes dart around as if watching an action film, the brain incapacitates the major skeletal muscles. Evolution must have favored this condition in early humans. Those who lacked it might have thrashed around and vocalized in their sleep, drawing the attention of predators or, worse, sleepwalking into harm's way.[11,12] But if your body un-paralyzes before waking up, you might wander out of the cave into the fangs of a sabertoothed tiger. Un-paralyze after you wake and, for a moment or two, you're locked into a deactivated body.

In addition to the regular stages, periods of semi-sleep occur at either end of the sleep cycle.[13] The *hypnagogic* phase[14] happens while zoning out into Stage 1. *Hypnopompic* sleep is the transition to wakefulness after REM. In both stages, elements of dreams can blend with waking perceptions.[15]

My dancing shirt experience was a great example of the hypnagogic phase. I was mostly conscious and aware of my surroundings, which is probably why I felt in control of things. But part of my brain was still producing dream-like visuals. Had I received other sensory information simultaneously, it could have integrated into the experience, or snapped me to full wakefulness. Had my mom implored me from the kitchen to get my butt out of bed, my dancing shirt might have spoken to me in her voice.

Scene Three. Strange experiences in semi-sleep are common and can range from delightful to terrifying. I'd been awake for a few minutes when I saw the old man's head in the cloud. I'd probably just completed a sleep cycle and was on my way to starting another. In my memory, my eyes were open while watching the head. In reality, conditions were ripe for me to close my eyes and slip unknowingly into a hypnopompic hallucination.

Not all semi-sleep experiences are interesting or pleasurable. You may believe you're awakening to the smell of smoke or a room on fire. War veterans with post-traumatic stress disorder[16] may be thrust into hypnopompic battles, only to discover they're paralyzed and can't defend themselves. Or worse, their sleep paralysis may lift before waking, compelling them to leap from the bed and lash out at imaginary foes. An incidental car horn becomes an air raid siren; a motorcycle backfiring becomes bomb strikes. Anyone else unlucky enough to be in the room becomes an enemy combatant.

Scene Four. Having a monster on my heels was no fun, and my poor young brain experienced an appropriate level of terror during the chase. What led to the abrupt self-awareness of both inhabiting a bad dream and having the power to flip the script? Was it a shortage of friends in my waking life? A cartoon I'd seen? An aversion to clichés? I don't know, but here's what I've learned.

It was a classic *lucid dream*, defined as becoming aware that you're dreaming while still engaged in the dream. Half of us experience them sometime in our lives, and a fifth of dreamers have them regularly.[17] Research summarized in *Psychology Today* identifies eight possible triggers of lucid dreaming.[18] Some involve exercises done while awake. For example, forming a habit during the day of asking "Am I dreaming this?" can increase the chances of asking the question while actually dreaming. Certain kinds of electrical brain stimulation, drugs, and sleep masks with flashing lights have also been shown to elicit lucid dreams.

In what sounds like the plot of a science fiction movie, lucidly dreaming subjects were taught to welcome researchers into their dreams in real-time. Subjects learned to respond to questions with eye movements or facial muscles during actual dreams.[19] In some cases, they'd hear the investigators' voices coming through a radio or spoken by another character, and the sleepers could respond to what they heard. They could also perform simple math problems and other small feats.[20] But the research wasn't just for curiosity's sake. Dreaming about an activity can enhance performance and creativity, so lucid dreaming could help people achieve their goals in real-life endeavors. Lucid dreams may also help people cope with trauma and chronic nightmares.

I still have the occasional lucid dream, but it's different now. They mostly happen when I dream I'm stuck in a boring situation or trying to find my way out of a maze of hallways or streets. I'll suddenly realize I'm dreaming and decide to leave the situation or simply wake up. It may not be as exciting as making friends with a monster but still qualifies as consciously extracting myself from an unpleasant situation.

Scene Five. My daughter doesn't recall having lucid dreams as a small child or being able to simply choose to step out of nightmares. But she did have a merciful memory. The specifics of her nightmares would often fade by the time she arrived at our bedside, making it easier to walk her back from whichever circle of hell she'd just returned. It's hard to stay frightened when you can't remember what frightened you.

As we age, we tend to become savvier about the differences between reality and fantasy. For kids, the distinctions aren't as sharp. Dreams for them can have a much greater impact on their emotions and temperament. Sadly, children with frequent nightmares can carry anxieties into their waking lives and dread having to go to bed at night potentially to confront yet another scary dream. Two or more nightmares a week may signal the need for professional help.[21]

My kid experienced nightmares at a typical rate, and we stumbled on effective treatments. The soothing back-to-bed escort was most common. More serious instances warranted more of a tag-team approach, combining two or more walks back to her bedroom. Escalation to the next level involved a snack or glass of warm milk in the kitchen, with a fun chat to push the bad dream further from memory.

In the long run, the most effective treatment was a stuffed pink bunny. His name was Joe. Full name, Joe Bunny. Joe was her go-to sleep partner into her late teens, a dedicated and selfless service bunny. He went everywhere with her, including sleep-away camps and family trips. When still in her early single digits, we explained that Joe Bunny gets scared when she gets scared, and that sometimes he has nightmares, too. When she believed he was frightened, she learned to provide him with soothing comfort, as we'd provided her. From there, it was a short leap for her to learn to calm and soothe herself. Nothing beats a good parenting hack.

Scene Six. My squash-pocalypse dream was a mild recurring nightmare I lived through several times. The majority of people have had at least one recurring dream. Mine has a backstory that explains it pretty well. Mom's go-to squash recipe yielded about eight servings of mashed, fluorescent-orange sludge. To my young taste buds, it was dry as sawdust, foul-smelling, and gag-triggering. I was not fond of it, but I was forbidden from leaving the table until I ate enough to cause permanent psychological damage. In other words, it was a real-life nightmare. Even then I understood that Mom only wanted to siphon some nutrition into her skinny, picky eater of a son. But knowing that wasn't enough to prevent a hated recurring side-dish from morphing into a recurring nightmare.

On the positive side, Mom's squash gifted me a perfect example of how dreams integrate elements of reality with fantasy. I knew implicitly what that orange substance was in the basin. I even knew it was about to explode. But I didn't know it was a dream until it ended with a bang and I was awake enough to realize I'd had *that* dream yet again.

I didn't try to interpret the squash dreams at the time. Later, in my teens, dream analysis became a fad among some of my friends. They were convinced by some pop psych books that our dreams were telling us things. They believed our dreams held coded advisories we ought to heed, or even offered insights into suppressed traumas or past lives we inhabited.

Sigmund Freud's 1899 book *The Interpretation of Dreams* had undue influence in legitimizing dream analysis. Variants are still practiced today by psychotherapists who believe dreams hold hidden meanings.[22] They contend that decoding dreams yields insights into the unconscious. Funny thing about hidden meanings: If you look for them, you'll think you're finding them everywhere. Folk singer Melanie Safka joked about this in one of her songs, declaring that anything longer than it is wide is a phallic symbol.

The same creative license that makes dream interpretation fun and fascinating is also its greatest liability. Clients and therapists are free to interpret the same dream in different ways. Maybe it's a good therapeutic device for starting conversations and getting troubled patients to open up, but proponents claim much more than that. Appropriately, there are more doubters than believers nowadays, and psychologists discredit the technique for its lack of scientific validity.[23]

Critics and proponents of dream interpretation agree on this much: Dreams incorporate bits and pieces of our everyday lives, as well as fears, joys, stresses, relationships, and desires we hold in our thoughts and memories.[24] As a general rule, it's best not to trust diagnostic conclusions inferred by dream analysis. One reason is that virtually everything in your dreams is longer than it is wide.

Scene Seven. Dreams can range from euphoric to horrifying. The most extremely awful end of the spectrum has a special name: *night terrors*. When I awoke from mine and my parents rushed in, I felt intense embarrassment on top of deathly fear. It seemed that I should have been able to tell my parents *something* about what had just happened to me instead of merely crying like a baby and saying, "I don't know." Vague as it was, I can still attest to its awfulness decades later and the depth of the feelings it evoked. And to this day, I still don't know what "it" was.

Night terrors aren't terribly rare at the age I had mine. According to sleep research, about one in twenty children experiences them.[25] There's seldom any danger for the children who suffer from them, and the bad dreams will stop on their own by the time they're teenagers. The incident I described was my one and only—but that's one more than I'd like to have had. It's got "terror" in the name for good reason.

Many books and articles have been published on sleep and dreams, penned by everyone from neuroscientists to self-proclaimed prophets. But as with most topics that enjoy widespread interest, some of these works make claims at the fringes of science—if not well beyond the fringes and completely over the edge.

Take *Zolar's Encyclopedia and Dictionary of Dreams (Fully Revised and Updated for the 21st Century)*.[26] Zolar offers over twenty thousand incredibly specific interpretations. For instance, did you dream of a married woman picking blackberries? If so, you'll soon become pregnant. (Beware, guys!) The author has enjoyed a long career using astrology and other paranormal methods to make claims that never get tested. He derives them from "mystic mathematics." I'm not a mystic mathematician, but it's safe to assume that this branch of math is unique in its freedom from the constraints of logic and validity.

Brave authors unflinchingly implore readers to believe that dreams convey valuable information from not only the stars but also from the future, dead humans, distant aliens, a collective consciousness, and our past lives.[27] The consensus response among scientists to such claims is a resounding "No, they don't." But this isn't because scientists are closed-minded. Many would love to be the first to offer unequivocal evidence for any of these extraordinary claims. The problem is that the evidence isn't there.

Another problem is that proponents on the paranormal fringe undervalue the brain's capacity as a wellspring of weird and self-deceptive experiences. So let's give credit where it's due. Let's talk brains.

The squishy organ inside your skull contains 170 billion microscopic cells, half of them *neurons*.[28] The power of neurons arises from a combination of their vast numbers and the dense web of interconnections among them.[29] One neuron can link to thousands of others, each of those to thousands more. So, when a neuron fires, its electrical "spark" can surge into a wave that activates millions of other neurons. This is nothing like how your home computer works. Still, it has inspired a modern type of *neural network* computer method that simulates neuronal processing—and human capabilities—with uncanny accuracy.[30]

It's borderline miraculous how a three-pound tangle of sparkers and wires manages everything from food digestion to self-awareness. But much of what once was miraculous is now merely amazing. The brain has given up a lot of secrets in recent years. We still don't fully understand exactly how webs of neurons give rise to complex effects such as a sense of consciousness.[31] But this gap in our knowledge is shrinking and still leaves the door open to a belief held by most people worldwide: *"I" am more than just neurons*.[32] But am I? Are you?

The belief that a mind, spirit, or soul exists separate from the body is called *dualism*. If it's true, those fringy ideas I mentioned earlier become more plausible. A soul unbound by the physical body could survive the body's death, reincarnate into a new person, or haunt our dreams and bedrooms.

My soul wants me to be a dualist. But I also know that just because I want something to be true, or even if it *really feels* true, doesn't make it so. The brain can generate the same experiences that dualists attribute to outside entities or forces. Today the best available evidence leans more heavily toward the *monist* side—the position that the brain alone creates the soul feeling and the feeling we're more than neurons. At least four streams of research persuade me that the monists make the better case.

First, illness and injury to the brain affect consciousness, self-awareness, personality, and memory.[33] The location and extent of the damage determine how much consciousness or memory is lost, how disheveled the sense of self becomes, or how the personality changes.

Second, bizarre experiences that seem real can be brought about electrically and chemically.[34,35] Hallucinogenic drug experiences align with the affected brain regions—visual, auditory, and so on. Recent work with controlled doses of LSD and Ecstasy (MDMA) suggests that some of these drugs may even repair damage that causes mood disorders.[36]

Third, fMRI scans (functional magnetic resonance imaging) produce live 3D maps where specific brain areas "light up" according to subjects' thoughts and experiences.[37,38]

Fourth, "deep learning" machines capable of *artificial intelligence* use neural network processors to mimic how human brains operate.[39,40] They are rapidly becoming more human-like and behave as though conscious and self-aware. They learn from mistakes, succeed at complex games such as chess, recognize faces, detect hidden patterns, translate across languages, and work around damaged or failed components. They also make obvious blunders, learn by experience, misperceive, misjudge, and misbehave. They show us how human intelligence and experience emerge from a mass of cells.

Dreams can feel convincingly dualistic, as if our souls have reached out and gathered information from parallel universes. Once, I dreamt that someone I didn't know told me a joke I'd never heard, complete from setup to punchline. I woke up laughing out loud. The experience was striking enough that I jotted down the joke before falling back to sleep. I woke up in the morning expecting to find I'd only written nonsense. I was surprised to find it was indeed a funny joke. I searched for it on the Internet to no avail. Although it absolutely felt like it had come to me from somewhere outside of myself, the simplest explanation was that it hadn't. Nothing about the joke suggested it couldn't have come from some part of my own mind. Also, it perfectly suited my sense of humor and arrived in the only language I speak fluently.[41]

My experiences were tame compared to the many bizarre sleep-related events reported across the centuries.[42] While we can't go back to confirm them, most sound like typical hallucinations generated by the brain during the hypnopompic and hypnagogic phases. What's striking is how different those descriptions often were compared to what we hear today. Hallucinations, it seems, follow trends.

Culture plays a big role in determining how we interpret experiences. Awaken in Rome around 400 A.D. and you might find yourself being molested by a male *incubus* or female *succubus*—a horny demon-like creature that only visits at night. If you're in Texas in the 1970s, the intimate invaders would more likely be a small group of gray, big-eyed, four-foot-tall, perverted extraterrestrials who enjoy inserting electronic devices into human orifices. In other times and places, it would be a witch or "hag" sitting on your chest, a demonic possession, or a close encounter with a neck-gnawing vampire. And yes, the nocturnal visitors can also be of the more welcome variety: beloved family members and pets, leprechauns, angels, or gods.

Sleep hallucinations and vivid dreams can feel as real as reality, transporting you into a fully formed world that feels authentic and completely outside of you, not at all shaped by your own mind. They feel real because they arise from the same parts of the brain that bring normal waking perceptions to consciousness. They hijack that system, substituting alternative content for regular sensory input.

What we experience in dreams needn't obey the usual physical laws. Consider what an amazing trick our brains play on us when they produce events in worlds that could never occur in reality *and*, during the dream, suppress whatever skepticism we'd have felt if awake. No wonder we sometimes awaken from a dream with a feeling that it *may* be possible some part of us just hovered above the sidewalk, hobnobbed with a celebrity, dunked a basketball, or spoke with a beloved relative lost to death years ago.

If our minds and dreams could access information that's normally unavailable—for example, from a remote location or the future—then we should be able to learn things that are both accurate and previously unknown to us. *Clairvoyance* means discerning things from immediate or remote environments that normally would be imperceptible. *Precognition* means doing so before it happens. My joke dream felt clairvoyant, but there's no reason my brain couldn't have used what it already knew—words, ideas, joke structures, etc.—to synthesize something original while asleep.[43] On the other hand, had I written out my dream joke in perfect Mandarin, I'd probably be a dualist today.

We take our cultures and subcultures for granted, even though they constrain our options and direct us to prefer certain foods, political parties, people, and religions. The social milieu into which we happen to be born furnishes the family, community, polity, education system, moral code, and historical times through which we do our best to navigate.

Cultural elements filter into day-to-day life, and day-to-day life filters into dreams. But like the fish taking for granted the water in which it swims, we tend not to realize how social environments shape us—and, by extension, shape our dreams. So while dreams come from the brain, the brain appropriates content from many sources around us. I cobbled together my childhood monster from various elements in my consciousness: movies, television, story motifs, hopes, and fears. So I agree with the biblical claim that "there is nothing new under the sun." Except I'd add "or the midnight sky" when it comes to dreams.

Our brains feel very busy. We process sensory information, experience consciousness and self-awareness, and have internal dialogues. We bring up memories, engage in complex social interactions, and dream. Despite our minds feeling rather preoccupied, we're oblivious to most of what's happening in our brains at any given time. Everything we think we know about reality—including ourselves—is only the tip of a neurological iceberg. The rest is underwater. Hard as we might try, most of it is inaccessible.[44]

The message is humbling: We don't know our minds as well as we believe we do. What we perceive as happening inside and outside our bodies is a story told to us by our brains as they try to create a sense of coherence and stability. Dreams are also stories authored by the brain—stories that routinely flout physical reality. Consistency with the physical world can tell us whether an experience we're having or recalling is real.

If you recall soaring above the landscape unaided by any devices, or stopping a monster in its tracks by saying hello to it, you can be fairly confident it was a dream.

Memories of dreams that don't overtly bend reality are a different matter. Sometimes we ask, in all sincerity, "Did that really happen, or did I dream it?" My mother's cognitive abilities declined steadily in the years before she passed at ninety-three. As her dementia progressed, she was increasingly unable to distinguish between dreams, memories, and real life. This could be jarring for anyone trying to talk with her. She'd wake up from a nap and say, "Oh, hi! I was just talking with my father upstairs." Except there was no upstairs, and her father was over fifty years in the grave. Fortunately, her brain's stories usually delighted her and rarely caused her any of the frustration and despair we might associate with dementia. So rather than try to correct her, family, friends, and caregivers learned to step into her fantasies and respond accordingly: "Oh, lovely! And how's Pa doing?"

I don't give dreams any unwarranted power. I'm as sure as anything that Mom wasn't *really* communicating with my grandfather's spirit. Nor do I believe dreams harbor advice or portend future events. But when my parents visit my dreams at least several times monthly, they make me think, feel, and remember things I might not otherwise experience. I like that.

Dreams can be random, complex, or beautifully structured stories. They may be like tedious slide shows or challenging quests. They can be anything. I see them as simply one of the great perks of being human.

CHAPTER THREE

Hanukkah Boys Wait Up for Santa

Yes, Virginia, there is a Santa Claus . . . your little friends are wrong. They have been affected by the skepticism of a skeptical age.
—Francis P. Church[1]

FRONTSTAGE

I'm the only Jewish kid out of thirty-two in my class. That makes me the closest thing to a minority most of them have ever known. I get teased sometimes, but everybody teases everybody about something, none of it very hurtful. It helps that I'm the tallest kid and mistakenly rumored to be one of the toughest first graders in the whole school.

Mrs. Garfield, my teacher whom I love, further elevates my status by devoting one of the holiday bulletin boards to Hanukkah. She puts me in charge of the design and works closely with me on the execution. That attention, and Hanukkah itself, so far are the only cultural benefits that accrue from being born into my religion. Three hours of Hebrew school every Sunday is my cross to bear.

There are three great things about celebrating Hanukkah. The first is that the gifting extends across eight evenings. Ignoring that there's only one gift per day and that gifts two through eight are usually lame, Jewish math calculates that Hanukkah is eight times better than Christmas.

Second, as the youngest, I usually get to light the menorah—the only time a six-year-old can work with open flames and not fear repercussion.

Third, Hanukkah arrives on different dates year to year, but on average, it beats Christmas by a couple weeks. Judaism thus reigns as the most enviable religion all the way up to the start of Christmas break. Then its star fades during the long wait for the holidays to return the next year.

To the other kids in class, the Hanukkah bulletin board is a tiny Jewish island in a sea of Christmas trappings extending beyond the classroom as far as the eye can see. The town is filthy with candy canes and lawn reindeer. The radio plays *only* Christmas songs. Homes and public spaces are done up in the colors, lights, and glitz of the season. There's a life-sized nativity scene at the church across the street from school. Passing it each day during the first three weeks of December, the discerning eye observes that two of the wise men keep switching positions. It's eerie. Turns out it's an annual prank by the big kids from the junior high. A few church ladies are up in arms about it, but Baby Jesus doesn't seem to mind. Elsewhere, grocery checkers, neighbors, dental hygienists, gas station attendants, waitresses, and everybody else greet you with a "Merry Christmas!" whether you like it or not.

I like it. People are nicer in December, and there's a good mood in the air. We get a break from school, and every kid gets caught up in the spirit of receiving.

One of the highlights of the season is when the family in the apartment next door invites my brother and me to help decorate their Christmas tree. Their two boys are our best friends, and it's always a blast. Barbara, their mom, plies us with sugar cookies while we busy ourselves unpacking box after box of ornaments, paper chains, popcorn strings, tinsel, and colored lights.

At the end of the evening, the tree sags under the weight of our garish accessorizing and we're all drunk on hot chocolate. But now the place is ready for Santa, and he will know exactly where to leave all the presents.

We prepare to return home. Small gifts are exchanged. Barbara helps us on with our coats and shoes and scarves and hats and mittens. We say our goodnights, and my brother and I stagger the ten feet across the porch from the neighbor's door to ours.

I can't *not* believe in Santa, given his presence everywhere except my family's apartment. I understand he's not Jewish and won't be coming down our chimney to deliver any gifts. But by Christmas Eve, my brother and I are already bored with our Hanukkah presents and primed for some vicarious comfort and joy. When the other neighborhood kids get presents, we *all* get presents. That's because by New Year's Day, all the new whiffle balls and bats, sleds, robots, Etch A Sketches, and LEGO sets become community property. We aren't excited for our friends. We're excited to *exploit* our friends.

This leads to our final Christmas ritual: Santa's arrival.

Our apartment is one of six in an old, converted house. We have front-facing windows, beneath which our living room couch is backed against the wall. We're convinced for no good reason that Santa's glide-path to the rooftop will take him straight across our field of view if we're looking out those windows. By seven o'clock, we're kneeling on the couch, our arms crossed on the backrest propping our chins, our eyes trained on the skies like NORAD. We await the sleigh, the reindeer, the man. This time we *will not* fall asleep.

I'm surprised that my parents abide such a pagan ritual. They've literally threatened to beat me if I so much as utter the words "Jesus Christ" under my breath, in vain or otherwise, like my dad does every day. "We don't believe in him" is the only explanation I receive. It'll be another year or two before they let slip that Santa isn't real and we don't believe in him either. Little did they realize that their one-eighty on such an important subject would establish in me a healthy skepticism toward all authorities—an attitude that more often than not serves me well for the rest of my life. But for now, they tolerate my delusion that a flying sleigh pulled by fancy deer and piloted by an obese saint will be landing on our shared roof at any moment.

Of course, we don't see Santa on that Christmas Eve of my first-grade winter break. One minute we're kneeling on the couch looking out the window, the next minute it's morning and I'm waking up in my bed. I remember it's Christmas Day. And there's the twinge of vicarious excitement about the presents other kids are opening.

Backstage

Ray Bradbury published *Dandelion Wine* in 1957.[2] It's a memory quilt of short stories about the adventures of twelve-year-old Douglas in his small-town summer of 1928. It brims with warmth and subtlety. I was only a teenager when I first read it, but it made me appreciate the beauty of simple, quiet things.

One of the stories opens with Douglas and his pal Charlie running to old Colonel Freeleigh's house. Charlie has promised Douglas there's a real time machine there. He wasn't lying. Soon after they arrive, the Colonel's words transport them back in time to the thunderous stampeding of bison on the American prairie; the performance of a legendary Chinese magician at a Boston theater; a soldier's-eye-view of the Civil War. You had to be back there to fully appreciate these things. Colonel Freeleigh was, and the boys felt like they were right there, and right then, with him.

Bradbury's short story stuck with me, but it was only recently that I personalized it: *I'm a time machine, too!* We all are. I want to advocate for active remembering and for not taking for granted that our brains evolved this ability. I want to remember the thrill of when my dad brought home my first two-wheeler. I also want to remember that it arrived heavily used, coated in rust, one speed only, with a brake lever you had to work by reaching down to the front fork and pushing it so it would scrape against the tire. I want to always remember my impatience when the first thing Dad did after bringing it home was to break it down, clean and lube the moving parts, sandpaper the entire frame, spray paint it red, and reassemble it back into a working bike. I want to always remember him teaching me to ride it the next day.

I'm not saying this to suggest that my nostalgia is more special than anyone else's. I just want to recommend you take trips in your time machine. At first, you'll see only the very best and the very worst memories. Be patient. Most of who you are is made of all those other memories in between. You'll find some.

Apropos of nothing, here's one I forgot that I remember. It's a brief scene at an amusement park. My family visited it exactly once, around 1960, making me about four years old. "Pleasure Island" existed from 1959 to 1969 in a Boston suburb. It was neither an island nor much pleasure. There was a long, sweaty drive in the summer heat in the non-air-conditioned Plymouth. Attractions were crowded, and there was little to do for those of us who were too small for the good rides.

The clearest memory I have of Pleasure Island is the "giant" treasure chest near the exit. It rested crookedly on some sand as though washed up from a sinking pirate ship. I'll guess it was about ten by fifteen feet and six feet tall. Its base was ringed with kids burrowing under the edge like ants. One by one, they'd disappear into the chest to investigate the treasure and crawl out beaming a minute later. My parents wouldn't let me dig, but my big brother was permitted. Stuck there watching, I was consumed with envy. That potent emotion is probably the only reason I'm able to recall this story.

Our rubber flip-flops went tacky on the hot pavement during what seemed like a twenty-five-mile trek back to the car, I asked my brother about the treasure chest. He said it was dark and empty inside, and it smelled like feet. There's probably a life lesson in there somewhere.

With the time machine metaphor and recollections like Pleasure Island, I've found that it's not only about the memories. It's also about the perspective from which we regard them. We tend to see a particular memory much the same way every time we pull it up into consciousness. But there's no law that we must. You can adopt different perspectives and see the same events differently through each.

My time machine has a switch that selects among three operating modes:

1. *Kid Mode.* I can see Christmas Eve 1962 as it lights up my retinas and tickles my eardrums. I see it unfold through a six-year-old's eyes and ears and brain. I re-feel the fun. I experience echoes of the laughter. I sense the anticipation, viscerally, that I felt just before we unboxed the tinsel and flung it up onto the tree, and onto the cat, and all over each other.

2. *Adult Mode.* I watch the same scene as though it's a documentary. The spareness of lower-middle-class furnishings, the big console television in the corner of the room, the absence of art on the walls. It's a scene of simple goodness. Secure and warm. It makes me happy now that *those* kids got to experience it then, oblivious as they were to the fact that one of them would be recalling it in the 2020s.

3. *Scientist Mode.* This perspective reveals an even wider spectrum of colors and facets. Maybe it's not for everyone, but I find it enriches the experience immensely. I get to see the hidden movement of the clockwork—how the machinery of cultural contexts, family dynamics, peer interactions, and child psychology colored my precious misunderstandings.

So it's back into the time machine as I switch it to Scientist Mode and have another look-in at the Hanukkah boys' Christmas.

Kids aren't born knowing how to do Christmas. *Socialization* is the lifelong process of learning one's group's norms and customs.[3] Children have to be socialized to believe what others accept as true and to behave as others expect them to.

It starts off for infants (and reiterates for adolescents) with a realization there's more in the universe than just "me." Understanding that others also have thoughts and feelings is a major leap. From there, we learn more about ourselves and others through everyday interactions and in more structured settings such as games and rituals. In this process, we gradually learn the "right" things to say and think.

Santa Claus, and all the other Christmas traditions, aren't cosmically preordained. They're *social constructions.* Evolved over centuries, Christmas has been molded by a million accumulated nudges from religious institutions, advertising agencies, communities, parents, and even children. But if all you know is your own traditions, they feel inevitable and eternal. Any variations from the theme cause discomfort and seem weird, or funny, or dangerous.

Kids are largely unaware that traditions vary widely across cultural groups. I recall when telling my little friends all about Hanukkah, a not uncommon response would be, "But you still have a Christmas tree, right?"

As soon as I was old enough to be socialized from sources outside my family, I learned as much about Christmas as anyone. Television was a big source. *A Charlie Brown Christmas*, *It's a Wonderful Life*, *Rudolph the Red-Nosed Reindeer*, Bing Crosby specials. The stores, schools, downtown streets, and everything else were festooned in red and green from here to the cultural horizon. I greeted people with a "Merry Christmas." Nobody said "Happy holidays." "Happy Hanukkah" was consigned to a few private households and the building and grounds of Temple Beth Shalom.

There are Hanukkah traditions, of course, though I'd wager that my neighbor family knew little or nothing about the rituals and customs of our biggest holiday. That was the norm in my town, and I didn't hold it against anyone. Hanukkah wasn't in the air *they* breathed as Christmas was in mine.

Most children's first socialization experiences are with the immediate family. But the family is not a remote island. Each has its little quirks, but most of the norms transmitted to the kids come from elsewhere. It's the delivery system for the ways of the larger culture. Many are functional tools for surviving in a complex and sometimes hostile world.

Norms and customs can be traced generationally, as when parents pass along to their children the same things they learned from their parents. But in modern societies, family socialization is heavily supplemented by information filtered through peers, extended families, mass media, religious affiliations, and the education system.

Groups don't exist in vacuums. They intersect. They nest within one another. They change and evolve from pressures emanating within and without. You've probably been socialized into multiple groups at different scales—from nation-sized to family-sized. Some norms and customs inevitably conflict. Your peer group may have different attitudes toward swearing than your church. Your parents may not approve of your being taught Darwinian evolution in school. Because norms may conflict, navigating them can be perilous. Not only do we all have to learn what to think and how to act, but we also must learn where and when different thoughts and actions are permissible. Otherwise we risk punishment and rejection. Society is a minefield.

Clashes and conflicts aren't the only possible outcomes when different groups bump up against each other. Good things happen, too. *Cultural diffusion* is the spread of innovations and traits through intergroup contacts.[4] It's how the American tradition of Christmas trees was imported from a German tradition which, in turn, evolved from nonreligious customs of bringing evergreen boughs inside during the long winter as a reminder of warmer, sunnier days to come.[5] Cultural diffusion also accounts for such things as Levi's in Latvia, crêpes in Cleveland, and K-pop in Kazakhstan. And it's why I know at least two Jewish families that actually did incorporate Christmas trees into their Hanukkah traditions.

When my first-grade teacher helped me put up the Hanukkah display on the bulletin board, a little bit of culture diffused. It was probably the first time most of my classmates were exposed to any sort of Jewish imagery and tradition.

When my brother and I went next door to help trim the tree, some Christian culture oozed into my family, too. We got a firsthand dose of that holiday above all holidays around which, as described in the film *A Christmas Story*, "the entire kid year revolved . . . [the] yearly bacchanalia of peace on earth and good will to men."[6] We had a firsthand experience that raised our awareness of what goes on in a Christian home to prepare for the holiday. Then we went home and told Mom and Dad everything we did and saw. So they learned, too.

Magical thinking is a belief in causal connections without evidence or plausibility. Examples are the belief that a rain dance brings rain or that a serial killer's car has cooties. Many magical beliefs are encouraged by religions. The real Christmas miracle may be the way the holiday converts a set of cultural traditions into a personal reality that feels as substantial as a fir tree dehydrating in the living room. For grown-ups, the absence of those traditions is unimaginable, so woven are they into the seasonal fabric of the family.

For small children, the flying reindeer and jolly gift-giver are as real as the tree, the cookies, and the presents. Children have magical thoughts because they trust what they're told by their socializers. Those thoughts become beliefs as they're reinforced by customs and rituals, by parents and clergy, and by platitudes like "Wishes can come true."

On the eve of that Christmas, kneeling on the couch and looking at the sky, my mind swirled with magical thoughts. I'm glad I had them. I'm also grateful that, later on, after "I put away childish things,"[7] I discovered the transportive power of my time machine, and the sciences that let me crack open those memories to reveal textures and colors I'd never have remembered on my own.

CHAPTER FOUR

You Sneeze, You Die

I asked God for a bike, but I know God doesn't work that way. So I stole a bike and asked for forgiveness.
—Emo Philips[1]

In one way, this chapter is no different from any other in the book. The Frontstage section recounts personal experiences with some extraordinary claims. The Backstage section examines them more closely through a skeptical lens.

In another way, this chapter differs from all the others in that it is the most likely to bother some readers. For this reason, I feel the need to preface it with some thoughts and clarifications before launching into it.

It tells the story of how a child—*this* child—drifted away from religion and stayed away. Hearing such things makes some people uncomfortable. If that's you, please skip the chapter. Really, it's okay. But please allow me to say just the following.

The main events of this chapter and my reactions to them are true. They happened. Extraordinary claims were made to me on behalf of the religion in which I was raised, and the doubts that emerged for me were major, formative events in my life. At the time, this little boy felt very alone with those doubts. I was made to feel this way by a number of well-meaning people who sincerely believed it would be better for me simply to accept the religious claims, which had triggered my doubts. That would have been my preference, too. But I couldn't talk myself into it. Instead, I continued to ask questions.

I was not a bad kid looking for trouble. I was actually a relatively good and obedient little boy, wanting to do the right thing. That included not lying about what I thought and felt.

Eventually, I came to understand that people far wiser than I had articulated reasons for their own lack of faith—reasons that made a lot more sense to me than those I was given to have faith. I'll talk about some of those reasons in this chapter, though it's not my purpose here to proselytize disbelief. If you are a person of faith, dear reader, that faith should not be threatened by reading about a child's experiences or by this adult's perspective on them. It's only my story, my take on things.

As a person of faith, you may find the chapter interesting for its insights into the mind of a child confronted with the same religious claims and experiences as yours but who took a different path. And you may find it interesting—even if misguided—to see me apply the same analytic approaches to my childhood religious experiences as those applied to nonreligious events in other chapters.

Or, as I've already suggested, you may skip ahead to read more about alleged psychics, ghosts, UFOs, and conspiracies. But as the author, I couldn't simply skip over such an important period of my life as though it didn't happen. Who isn't deeply affected by what happens to them between the ages of six and nine?

So let's join little Barry on his first day of school.

FRONTSTAGE

"Good morning, class!"

First words, first teacher, first day, first grade.

She continues with sing-songy positivity and precision: "Now *you* say, 'Good morning, Mrs. Garfield'! Ready? Good morning, class!"

"Good *morn*ing Mrs. Gar*field*," we reply in unison, mimicking her chirpy tone. We desperately want to please her. We're all nervous, and a few are still whimpering after their mom's goodbyes were said only a few minutes ago.

She goes on in that inimitable teacherly tone: "Welcome to the 1962–1963 school year! How many of you know The Lord's Prayer?

Raise your *right* hand if you do." She raises her own right hand and looks around the room.

Eventually, most of the "yes" kids resolve the left/right problem, and around three-quarters of the hands go up.

"Okay. If you know it, please recite it with me. If you don't, try to learn the first line today. It starts 'Our father who art in heaven, hallowed be thy name.' Can you all repeat that for me?"

Some of those words are foreign to me. What comes out is something like, "Our father whose art's in heaven, 'Hello' be they name." But my verbal gaffes go unheard beneath the collective mimicry of thirty-two other timid students. By next week, I'll have the whole thing down by heart.

I want to be a *good* student. Apparently, so do all the others. As the fall progresses, I hear plenty of grousing about all the reading, writing, and 'rithmetic expected of us. But no one ever complains about starting each day with Mrs. G.'s Lord's Prayer.

The morning prayer becomes a routine, like the Pledge of Allegiance, and in time, I think literally nothing of either recitation. Neither do I think about religion during the school day. Mrs. Garfield never pushes it and never even talks about it. What she does talk about are the Three Rs and how to be a good person. Mrs. G. was the best.

I pray aloud every morning in class until the middle of the school year when, for some reason, Mrs. G. stops leading us in prayer. But I dutifully continue on my own, as instructed by my parents. Every night I kneel at my bedside, hands clasped, softly reciting "Now I lay me down to sleep . . ." and hoping not to die before I wake. Works every time.

I pray for my family, and for most of 1964, I pray that they catch the Boston Strangler, whose murder spree has now reached into the suburbs. It's great local news fodder, but it scares the parents into a state of paranoia, which then scares all the kids. When my prayers are answered and they catch the creep in October, I feel civic pride for having played a prayerful role in his capture.

For most of the year, I also pray for a guitar. I've fantasized about *being* John Lennon since February when the Beatles burst onto *The Ed*

Sullivan Show and hooked me. But the holidays come and go, and still no guitar. Duly noted, God.

The wonderful Mrs. G set me up well for second grade. She taught us the alphabet and the numbers zero to ten. Later she taught us words, sentences, simple arithmetic, and how to write characters on yellow lined paper with a no. 2 pencil. I learned to work and play with others, some basic geography, and rudimentary social studies. I made more progress in that nine-month period than in any other since my time in utero.

Next fall, my second-grade teacher, three rooms down from Mrs. Garfield, is Mrs. Norton. I like her a lot, too. One Friday during the first month of the new school year, there's a substitute in Mrs. G's Room 101. If Mrs. Norton knows what's up with my old teacher, she doesn't let on. I go about my usual school-day business, head home, have some cookies and milk, play outside, and come in for dinner. No playing allowed outside after dinner on school nights, so I settle into my spot on the living room rug with an Erector Set project. Dad's nearby in his easy chair reading the late edition of the local paper. Mom is sewing something, seated in her chair across the wide lamp table from Dad. Older brother Dave is doing homework on the couch.

"Jesus Christ!" Dad says. I assume he's upset about something on the sports page. Without looking up from the paper, and to no one in particular, he announces, "A teacher at Memorial School was killed yesterday."

The previous day, Stanley Garfield, eighteen years old, arrived home after going AWOL from the Navy. He waited for his mother to return from her teaching job at the school. My former teacher walked into her home. Then her son, Stanley, strangled her to death. Soon afterward, Mrs. G's nine-year-old son got home from school and met the same fate. Later, when Mrs. G.'s husband, Robert, arrived home from work, Stanley shot him in the chest with a 22-caliber rifle. Stanley then fled despite his dad's pleas to call for help. Robert survived.[2,3]

My Mom gets up and sits next to Dad on the arm of his chair. They scan the article together. A few quiet words pass between them, probably something like "What do we tell the boys?" and "I don't know . . . we have to say *something*." My brother and I watch, listen, and wait for more information. Dad prepares to take charge.

"Boys, come over here." His tone is more like a question than a command. We go over and sit cross-legged on the floor near his feet. He looks at us gravely and says, "This is really terrible news. Mrs. Garfield was killed yesterday by her son. Mrs. Garfield is with God now."

My brother shouts, "No! Nooooo!" He's a fourth grader at Memorial, but Mrs. G. was also his first-grade teacher three years ago, and he still talks about her. His keening continues, and my parents don't know what to say or do for him.

I can't fully process the news. I haven't seen Mrs. G. in over three months, which seems like a very long time. I feel kind of sad, but it's sad like when a friend moves away. My seven-year-old brain's inexperience with death largely protects me from serious shock or grief. My parents check with me.

"Are you okay?" Mom asks.

"I'm okay," I reply. "I think God is letting Mrs. G. into heaven because she prays to him every day. And now I'll pray for her, too, to make sure she gets in."

Every kid I know at school and in the neighborhood gets religious training on Sunday at their church or synagogue. Catholic, Protestant, and Jewish are the only games in town. If other religions are represented, they're laying low.

Family, teachers, and temple officiates drive home to me and my clan that, of the three religions, we Jews have always been the most persecuted, and we've out-suffered the other two religions by wide margins. It's also taught that our people are the Chosen Ones, blessed by God above all others. We're special, and we're proud—which is okay because pride is not an official sin for us.

We're encouraged to recognize the slightest hints of antisemitism. I'm praised for recounting an incident where an older kid in a pickup wiffleball game at the park looked me right in the eye and sneered, "Jew-boy!" I happened to be the pitcher in that game. Next time he was up to bat, I hit him square in the face with a knuckleball. *An eye for an eye*, I thought. *It's in the Bible. Don't mess with us.*

We're also taught that our beliefs, and only our beliefs, about God, history, mankind, and the universe are true. This is reinforced in my Hebrew School at Temple Beth Shalom. We learn the rudiments of the language and history of the Jewish culture. We go down the Greatest Hits list of prayers and learn to recite them in both English and phonetic Hebrew. We learn the long history of Israel, which was finally established fourteen years ago as a sovereign nation. We learn about our special holidays that Christians don't get to have.

Most of us misbehave a little more in Hebrew school than over in our public schools. Hebrew school doesn't feel like *real* school. One of the reasons for this is that the teachers hold us to lower standards of discipline. Another is the storytelling.

The part of Hebrew school that's nearest to fun is the Bible stories. But my teachers and temple leaders don't call them "stories." They call them "history," as recorded in the Old Testament, and retold to us by serious men in dark clothing. Usually it's our regular teacher, Mr. Goldman. Sometimes our resident Rabbi Rudanski or the Reverend Beecher visits class to share stories.

The stories are wild. A guy stuck in a big fish's belly. A naked guy in a garden with a naked lady and a talking snake. A little guy kills a really big guy in a fight. Another guy saves the world from a flood by building a big boat for all the animals and his family. A guy almost kills his son because God tells him to, except God says, "Just kidding!" at the last second. One of the stories is like a disaster movie with attacks from insects and frogs, and with days of darkness and boils and baby-killings.

Our Methuselan Reverend Beecher tells these stories animatedly, like an eyewitness. Being as old as the planet means he's a de facto elder of the temple. Besides him, there are no official elders at Beth Shalom, nor "Reverends" for that matter. We all love him. We just don't really know what he *is*.

But we know what he *does*. He's a utility player. He's asked to play a variety of roles at temple services. He subs when regular teachers are absent. And he offers a kindly ear to any student or parent that might wish to talk with him about faith, temple policies, interpreting scripture, or personal issues.

One of Mr. Beecher's stories is about why we say "God bless you" when someone sneezes.

"Back in ancient times," he explains, "few people lived long and healthy lives as most do today. There were wars, plagues, mothers and babies dying in childbirth, and suicides."

Perhaps a bit much to lay on innocent seven-year-olds. Then again, we're exposed to Old Testament horrors far worse than these. We give Mr. Beecher wide latitude with his stories. He continues.

"Sneezing was taken very seriously because, in those days, if you sneezed, you died."

The Reverend pauses for effect. I'm looking around the room and see most of the other students hanging on his every word. I'd heard about the other causes of death, but sneezing is a new one for me. Next to me sits my best friend Alan. We look at each other at the same time and simultaneously mouth "Whah??"

Apparently at least one other classmate is also surprised by the sneeze-equals-death claim. She raises her hand to question it.

"Really?" asks Melissa when called upon. "A sneeze would kill you?"

"Yes, it's true," replies the kindly Reverend, "unless there was someone present to offer a blessing. If the person who sneezed received an immediate 'God bless you,' they'd be alright. They would survive. And so the tradition of blessing someone after they sneeze continues to this day."

Alan and I take the Reverend's claim seriously enough to discuss it in scrawled notes passed back and forth. We talk about it again later after school lets out while we wait for our dads to pick us up outside the building. On the ride home, I ask Dad if Mr. Beecher's sneezing story is true. All he says is, "Well, if Mr. Beecher says it's true, then it's true."

Maybe it's because I hadn't yet received the guitar for which I'd prayed so fervently. Maybe it's because of the end of our school prayers, or the end of Mrs. Garfield, followed ten weeks later by the end of President Kennedy. Or maybe it was the Sneeze of Death story. My relationship with God is approaching a critical juncture, and I start to wonder where my prayers are going.

BACKSTAGE

Mrs. Garfield quit leading us in Our Fathers when the U.S. Supreme Court prohibited organized prayer in public schools.[4] The court saw the practice as endorsing religion, crossing the U.S. Constitution's line separating church and state. Two-thirds of Americans disagreed with the ruling.[5] Banning school prayer, they felt, was not at all neutral toward religion. Most felt it was *anti*-religion, and to this day, a majority of the country agrees.

The court's argument for its decision has always been less well-known than the decision itself. As its supporting arguments made clear, the First Amendment to the Constitution protects religions from interference by the government and from favoritism toward any particular religion. Typically, prayers conducted in classrooms, before sporting events, at graduation ceremonies, or before school board meetings, favored Christianity above other religions. Mrs. G.'s Lord's Prayer, for example, came word-for-word from the Gospel According to Matthew in the New Testament of the Holy Bible (6:9–15). A fun thought experiment would be to imagine how the majority of the public would react to a reversal of the Supreme Court's decision about mandated prayers but some days requiring them to come from the Hebrew Bible, other days from the Muslim Koran, The Bhagavad Gita favored by the Hare Krishnas, and so on.

Another popular belief is that the 1962 Supreme Court decision outlawed religion and prayer in public schools. It didn't. It outlawed religious activities organized by agents of the state—teachers, administrators, school boards, and so on. Kids could pray on their own, organize prayer groups at recess, use school facilities for after-hours praying, and so on. Teachers could also still teach about religions. They just couldn't promote them.

For me and most other kids, the end of organized school prayer didn't seem to affect our religious feelings one way or the other. The morning school prayer had become a mindless ritual, something to be gotten through—the same way families sprint through grace before meals. *Good food, good meat, good God, let's eat!*

Death was a stranger to me before second grade. I'd not yet lost a pet. No grandparents passed since my birth. Mrs. G. was the first living creature personally known to me to slip the surly bonds of earth. I knew heaven is where good people go when God takes them home and that it's fabulous. My brother knew this, too, so I didn't really understand his over-the-top grief over Mrs. Garfield's murder or my parents' grim demeanors. If she were in a better place, shouldn't we be happy for her?

The outpouring over President Kennedy's assassination made a little more sense. At the time, my understanding of the political sphere, on a scale from zero to ten, was nil. But the president was in the local and national news practically every day of my life. He grew up and went to college around twenty miles from where I lived. He talked like us. When I was born, he'd already risen from U.S. Congressman to U.S. Senator, and then he was elected to the presidency when I was four. I knew little of what that all meant, but I had a firm grasp of the excitement and pride surrounding it.

The collective grief following JFK's assassination was palpable. I watched the funeral live on TV. I saw his family mourn and pray, including his two little children, both close to my age. I heard over and over about the thoughts and prayers offered by a heartbroken nation on their behalf. I prayed, too.

But questions continued bubbling up. I wondered how God fit into any of this. If he's all-powerful, what use did he have for my teacher and my president? And if he already had a plan and knew what he was going to do, what difference would my prayer make?

As time went on, my questions multiplied, but I never really got any good answers. Mr. Beecher, Mom and Dad, my rabbi, other temple authorities, and even some of my friends discouraged me from asking. In lieu of answers and explanations, I was usually admonished and told to "just have faith."

"Why should I have faith?" I'd ask.

"Because those who do are rewarded," I was told.

JFK and Mrs. G. were persons of faith. I sure didn't want to be rewarded like they were. And it didn't make sense to me that having faith was like buying an insurance policy for a happy afterlife. Being good people should be the point—not because it guarantees rewards but because it's best for everyone.

Mr. Beecher's Sneeze of Death story was the straw that broke my faith's back. Two aspects of the claim left me incredulous.

First was the idea that a sneeze would kill you. It sounded like a whopper of a tall tale, ridiculous on the face of it. The Reverend should have offered a more plausible explanation. I've since read that sneezing was a symptom of the onset of bubonic plague, and catching the plague meant certain death. So people said "God bless you" in hopes it would protect you. Most didn't die because they didn't actually have the plague. I wouldn't have doubted that story.[6]

The second issue had more to do with practical questions, captured in a series of "what-ifs."

Since everybody probably sneezed at some time, *what if* a sneeze burst forth while away from other people, say, lying in bed alone at night? Would you have to run around and find someone to rouse from a sound sleep and groggily bless you to save your life?

What if that took too long? How much time did you have after a sneeze for the blessing to be administered?

What if you blessed yourself? Would "Bless me!" do the job? If not, *why* not?

Would it count if the blesser was from a different religion?

Thinking about Mr. Beecher's extraordinary claim today, it's fun to push the questions even further. In the hypothetical world where a sneeze is sure death, could a rabbi make a blanket blessing in front of his congregation, perhaps covering each member's next sneeze? Or their next ten?

What about "sneeze attacks" like the ones I used to get from hay fever in the spring? I might have seven sneezes in rapid succession, with virtually no time for blessings in between. So given that sneezes one through six obviously didn't kill me, would I still need to get blessed after seven? Would I require retrospective God-bless-you's for one through six?

What if my pet sneezes? Dogs and cats do so commonly. Do the same rules apply? Would the blesser have to be human? Can animals bless each other?

Just one more. What if you're a baby, or comatose? Neither has chosen to be in their condition. Sneezing is a reflex response that attempts to clear irritants from nasal passages, and they're known to spew forth from both babies and the comatose. *Why,* I imagine asking the God of Abraham, Isaac, and Jacob, *if you exist, and if people are your creation, did you endow us with an involuntary response that would kill us if we're not blessed in time? A little capricious, no?*

If the highly respected Reverend Beecher stretched the truth, and the Bible tales were just as unbelievable, where did that leave me? I had no quarrel with the life lessons embedded inside the Bible stories, delivered as they were with whimsy and drama. I was already swimming in the sea of Jewish culture and taking my place there for granted. But with my faith teetering, God's stories rang no truer for me than the fairytales in books written for younger minds.

Then there was science. In contrast to Hebrew school, my regular school gave me plausible answers to my questions about the world. Explanations were coherent, broad, deep, and backed by evidence. I didn't feel worshipful toward science, but it surely evoked feelings of awe and humility in me. And it did so without demanding my loyalty or threatening my soul if I stray.

Gradually, I came to grips with a vague notion that, years later, I found perfectly expressed by Carl Sagan in his book *Broca's Brain*:[7] "As we learn more and more about the universe, there seems less and less for God to do."

I didn't try to hide my growing doubts, but neither did I go around blabbing about them. It was scary. My parents knew and tried to shame me. My Hebrew school teachers suspected, and I gradually disengaged. I truly didn't want to go against the grain, against these influential authority figures in my life. But the process was as involuntary as sneezing. I must have wondered why, if God existed, had He endowed me with a doubting mind. Was it a test? Was skepticism being laid before me as a temptation to overcome? Why the games? Why not just create me with a believing brain?

I saw no alternative but the existence of God being an open question. That made me agnostic about the God of my family's religion. As for other religions' gods, I didn't really think about them. If pressed, I'd have probably admitted that I was an atheist when it came to Baal, Thor, Eros, and the roughly ten thousand other gods that have been or are still being worshipped.

I stopped praying.

The grown-ups warned me about "The Atheists" and only spoke of them with disdain. From what I could infer, they were an evil cult that went around evangelizing disbelief and recruiting impressionable kids. Devil worship may have been mentioned. Miraculously, I'd arrived at my agnosticism quite independently. I didn't know any atheists or even agnostics. I didn't know that such people existed—at least not until I was warned against becoming one of them.

When I discreetly expressed my doubts about religion to a grown-up friend of the family, he said the smartest thing was to keep being faithful. If the Bible's right and infinite heavenly rewards await, he explained, my faith is a small price to pay to not miss out.

This is called Pascal's Wager, though neither of us knew that at the time. It's the idea that if you have to bet on whether God exists or not, you've nothing to lose and everything to gain by going with God. If you're wrong, you'll just turn to dust like everyone else. But if you're right, you'll win the mother of all lotteries.

"Not so fast," say the philosophers.[8] They point to a number of issues with the argument behind the Wager.

First, a doubter's faith can't simply be flipped on at will in response to an argument that she ought to believe. One can choose to stack the deck and selectively expose oneself to religious influences. But that doesn't guarantee true faith will emerge. A case in point: I was exposed to all manner of religious influences, including friends, family, authority figures, and culture. I *wanted* to believe. It didn't work. You can lead a horse to holy water, but you can't make him drink it.

Second, Pascal's Wager has been interpreted as a case for faking it. Since you can't simply decide to believe, you can at least "go with God" (or gods, or goddesses) and live your life as a good Christian, Jew, Muslim, Hindu, Buddhist, Sikh, and so on. Fingers crossed the chosen god isn't omniscient and won't see through your little deception.

This raises a third issue. Since there's an array of gods from which to choose, what if you go with the wrong one? The result would be a lifelong commitment to religious thought and action wasted on a false deity. That alone would seem to flip the odds on Pascal's Wager.

Then there's this: Why would a deity care about a person's beliefs above their inherent goodness, their moral character, and their treatment of others? What if one lives a good and moral life but doesn't happen to subscribe to supernatural beings?[9] Their only sin is lumping the "right" god in with the thousands of others they don't believe in. If morality and goodness are the goals, does such a person deserve exclusion from heavenly rewards, or worse, being sentenced to eternal damnation?

As I said, I wished I believed in God, especially as a child. It would have meant not disappointing my parents and maybe fitting in better socially. I wouldn't have needed to wonder what was wrong with me. Why couldn't my mind simply trust that God had a plan that didn't necessarily have to make any sense to me? If, as told, I was created in God's image, how could I turn out so wrongheaded?

Today, if I had to bet all or nothing, I'd come down on the side of disbelief. But I'd be a fool to say that I'm 100 percent sure there is no god. So in case I'm wrong, here's my prayer:

Dear god/gods/goddess/goddesses,

If you are reading this, and if you are omniscient, then you already know what I'm about to say: Though I was wrong, I came to my disbelief about you in full sincerity. I have used to the very best of my abilities the brain with which you endowed me. And at least since the chaos of adolescence (was that whole phase really necessary?), I have always tried to be good and to do good without you. If it was you and not I who caused me to strive to be a good person, then I'm sorry not to have thanked you and credited you sooner.

Thank you! And please don't hurt me.

CHAPTER FIVE

U.F. . . . Ohhh?

Not only do we live among the stars, the stars live within us.
—Neil deGrasse Tyson[1]

FRONTSTAGE
Something up in the evening sky makes no sense.

We're twelve years old in the summer of 1968. By day, we're charter members of a gang of boys called the JABETs: Jeff, Andy, Barry, Eddie, and Terry. We are naïve, prepubescent suburban tweens who wouldn't know how to get into serious trouble if we wanted to.

Our parents let us mostly run wild. Well, run *mild,* if I'm being honest. We've never been in serious trouble, and we've learned to make our own fun. In exchange for keeping out of the parents' hair until mealtime or dark, they leave us alone.

That feeling of independence pumps our creative juices. We set up obstacle courses for our bikes in backyards. In the woods, we erect huge, spongy "Indian mounds" of pine needles for no reason other than to take running, flying, kawabunga dives into them. We fashion working skateboards out of found wooden planks and siblings' outgrown roller skates. We shoehorn baseball diamonds into wherever they'll fit and play daylong, fifty-five-inning games.

We build forts.

Recently my dad tore down fifty feet of a rotting white picket fence bordering a neighbor's yard. He piled six-foot sections in the far back corner of the yard behind some shrubbery. For the JABATs, it's like having our own building supply store. Where Dad saw trash, we see opportunity. We put our heads together and plan a construction project.

Day One. We borrow hammers from our dads' toolboxes and pull nails from the pickets. After a few hours' work, we have neat piles of rails, planks, and rusty nails. As the demolition phase comes to a close, Terry kneels across a stray picket and almost ruins the whole project. He impales his shin with the business end of a protruding nail. His wailing could be heard half a block away in all directions. He's frozen with fear in an awkward half-kneel. We're frozen, too, because we're grossed out and afraid to grab either the plank or the shin and yank the nail out of him.

Somebody's mom calls the police. A small crowd of family and friends gathers. A kind officer arrives, assesses the situation, distracts Terry for a second, and jerks the board away from the leg, taking the nail with it. A small trickle of blood drips from the tiny wound. The cop scoops him up, hustles over to the cruiser, and drives off to the emergency room without lights or siren.

Day Two. Terry's back wearing a Band-Aid and a nonchalant attitude. Given the minor and freakish nature of the accident, the authorities (the moms) allow the project to continue.

Assembly is efficient. We renail the rails and pickets into lattice-like panels. We connect the panels to frame a large box for the walls and roof, like a shabby-looking modular home. We cover the sides and floor with musty sheets and blankets from our garages. A painting tarp waterproofs the roof, held in place with rocks at the four corners. The best fort we've ever made is done before any of us are called home for dinner.

It's an Alcatraz sweatbox and the Palace of Versailles rolled into one. Best of all, it's *ours*. It easily accommodates all five of us lying down. We make it homey with shelves and nail-hooks stocked with treasures like flashlights, good luck charms, baseball caps, and comic books. Somebody drags in a cooler. We stock it with extra sodas before realizing it's not actually a refrigerator. But my house is twenty feet away and, over the complaints of my mother, we slink in and abscond with aluminum trays of ice cubes as quickly as the old G.E. can produce them.

Day Three. Someone offers the idea of us all camping out overnight in the fort. My family and I, we're not campers. In twelve summers of opportunity, the closest I've come were humid nights in a YMCA overnight camp cabin, and a few more in a patio lounge chair on our screened-in back porch. I know it'll be an uphill fight to get permission from my parents to let us bivouac in Fort Knox. (We're calling it "Fort Knox" now.) We fan out to our respective homes.

Let the negotiations begin. "Mom, did you get your hair done?" I ask for the first time ever. "Looks nice!"

"No. What do you want?" she replies with weary efficiency.

"Nothin'. Errmm. . . . It's just some of the guys are gonna camp out in the fort tonight," I white lie, as if it's already a done deal. "Can I, too?"

"No." I fully expected her reply but planned several moves ahead in this chess game. I unveil the first:

"Awww, Mom, how come?"

"Because I said so." Also anticipated. Next move:

"MOM! My friends are gonna be right out there in the back yard tonight. Come on. PLEEEEASE?" How she could possibly resist the logical force of this argument was beyond me.

"Wait 'til your father gets home." This is her standard conversation stopper. Still, it's a good step up from "NO!" I'm halfway there.

My dad pulls into the driveway at 5:30 as always. He walks in the back door, and I'm on him before he hangs up his cap.

"Dad, can I sleep out in the fort tonight with the guys?"

"I said 'NO!,'" my mom's voice shoves its way in from the kitchen.

But Dad proves to be far more rational than Mom. Not missing a beat, he says, "Mow the lawn the next four Saturdays? Front and back?"

"Oh. Hmm. Okay." We have a filthy, greasy, smelly, teeth-rattling, refurbished, self-propelled, antique gas mower that sounds like a propeller plane taking off. It's magnificent. I *love* it when Dad lets me use it and feel sneaky pretending this is a concession. (In retrospect, I'm sure he knew this.) I doubt his heart is ever in it when he comes home from work and is immediately assigned by Mom to follow through with negotiations or punishments that he had nothing to do with.

"And straighten up your damn room!"

This was the real price for sleep-out permission. Painful but doable. So I agree to pay it. "Deal."

"Deal," he replies.

I hesitate for a second, unprepared for the fast compromise. Before he can change his mind, I'm charging out the screen door yelling, "Thanks!" The door slams with a *thwack*, and I'm racing to the neighbor kids' houses to see who else might have succeeded in their negotiations. Turns out only Jeff, but that's enough. We pack duffle bags with everything we, and a family of eleven, might need for the overnight.

That night, under the stars, Jeff and I talk quite a while about how cool it is to be out camping. There's only so much to say about that, so we move on to the next item on the agenda. Baseball. Our feelings toward the Boston Red Sox are the closest either of us has yet come to experiencing romance. Last fall they fell short of winning the World Series by one horrible seventh game. It's up to us to figure out why, already a third of the way through the current season, they've won only half their games.

From there, the conversations skip around aimlessly. We weigh the validity of a rumor that a bully from the next neighborhood was struck by lightning.

We name the names of the girls we hate most, which is to say, the ones we have crushes on.

We try to process why Martin Luther King was murdered just weeks ago.

We argue passionately about which sandwich is best: baloney and cheese, or fluffernutter.

By the time the lights go out over in my house, we're starting to get a little bored. We spread a blanket next to Fort Knox and lay on our backs looking for shooting stars. There's no air or light pollution in the far exurbs of Boston in 1968. My first-ever camping experience is infinitely clear and moonless. We spot several tell-tale streaks of light, reminding us that the sky is alive.

After a few more vigilant moments, Jeff breaks the silence. A bit of excitement in his voice, he asks, "What do you think *that* is?" He's pointing up at the sky, but I don't know that because it's pitch dark.

"Where?" I'm suspicious he's joking.

"Watch!" He feels for my arm and grabs it in two places. "Point your finger." Then he aims my extended arm toward the sky and sights it like a rifle. "Now look there," he instructs.

"What are you talking . . . ohhhhhhh. . . ." My voice trails off as I see what he sees. It's a moving point of light, clearly brighter than the background stars. Not Venus-bright, not shooting fast like a meteor, but prominent enough. Being a space nerd, I venture, "A satellite?"

Then the thing happens that we can't explain. The light slows to a brief stop and then reverses direction. We joke about it being a UFO as though we believe it isn't because it's so unspectacular. But we're baffled. Really baffled.

"Maybe it's looking for a place to land," Jeff speculates, trying to make it more interesting.

"Maybe it's looking for *us*," I say, playing along.

The conversation switches to alien invasions as we watch it cycle back and forth about five times in five minutes. But nothing else happens, so we're losing interest.

We enter the fort, snatch some comic books and flashlights off the walls, and settle in. We're still chattering more than reading. Not exactly the high-level philosophical discourse we believe it to be, but interesting. We play with ideas about life elsewhere in the universe, the chances we'll ever encounter it, and what it would mean if we did. Not "we" in the sense of humankind. "We" as in Jeff and me.

Part of me knew it wasn't just about us, though. We're both humbled by the huge sky and the boisterous night sounds coming from the woods not ten feet away at the boundary of the small backyard. As though establishing our own church to foster comfort and a sense of significance, we close the evening with a spontaneous call-and-response about extraterrestrial life.

"I wonder what kind of fuel they use," Jeff begins.

"I wonder if they have girls," I contribute.

"I wonder where else they've been."

"I wonder if they have sports."

"I wonder if they're friendly."

"I wonder if they have music."

"I wonder what they've learned?" says Jeff, triggering his drifting imagination and derailing him from the conversation. He's done.

"I wonder if they wonder about us," I reflect sleepily, the last words between us until morning.

Backstage

It was not so long ago—1947 to be precise—that newspapers came up with the label "flying saucer." A private pilot named Kenneth Arnold was flying over Mineral, Washington, near Mount Rainier when he saw what he believed to be nine shiny, disk-shaped flying objects. They gave him "an eerie feeling." He showed up at the offices of the *East Oregonian* newspaper two hundred miles away and told his story. They ran the flying saucer story because he seemed like a reliable witness. The Associated Press picked it up, and it went the mid-twentieth-century version of viral.

That same year, tiny Roswell, New Mexico, inadvertently thrust its way into the national psyche. First, as the location of another reported sighting. Second, as being seventy-five miles from the debris field of an alleged UFO crash. Third, as the rumored recovery location for the bodies of several alien beings. Fourth, as the centerpiece of a massive cover-up of these events by the US government.

Investigations later found the "crashed saucer" to be pieces of a balloon array carrying experimental military devices. The alien bodies claim is highly disputable, based on questionable testimonials and no hard evidence. However, as if to validate the old joke that "military intelligence" is an oxymoron, the Roswell Army Air Field first claimed to have recovered a flying saucer, then retracted their statement, and then issued another saying it was a weather balloon. Rumors and conspiracy theories about airborne dinnerware took metaphorical flight, unconcerned about the woeful lack of decent evidence for extraterrestrial spacecraft and pilots.

In the years that followed, blurry images of UFOs appeared everywhere. Tales of saucer encounters and abductions by their occupants became news and entertainment, and the line between the two increasingly blurred. Blockbuster films, best-selling books, late-night radio shows, TV programs, and eventually popular websites all riffed on these themes, keeping alive hopes and fears that "They" were always just over the horizon.

A lot of us kids were sky-oriented in the 1960s. We took the word of astronomers who, earlier in the century, squinted through crude telescopes and believed they could see canals on Mars. We saw their drawings on TV and in books, seemingly unassailable evidence of an alien civilization. Somehow we missed the news that better telescopes and several recent Mars probes proved there were no canals.

I was as obsessed with UFOs as I was with bottle rockets, war games, and monster movies. But even before I read any skeptical literature, I was growing suspicious that news of sightings and encounters may have been overblown.

It always struck me how most of the pictures of UFOs were terrible. As for the better ones, why were there never two or more of the same spacecraft? A shiny metal ship with beaming, colored lights is hovering over the town, an event of unimaginable historical significance. But every single time, there's only ever exactly one person who snaps exactly one photo of it on their seventeen-dollar Kodak Instamatic?

Robert Sheaffer called this the "shyness effect"[2] and illustrated its absurdity with a real event. On August 10, 1972, an unexpected, short-lived object from outer space appeared over sparsely populated areas of the western United States and Canada. Despite the suddenness, brevity, old tech, and population sparseness, there was extensive photographic and motion picture evidence from different vantage points, allowing the UFO to be identified. It was a meteor.

But the UFO I watched in the sky on that summer night of my tweenhood was no meteor.[3] It was, is, and probably always will be unidentified. I say this with some confidence because even one of the world's top skeptical UFO experts, Mick West, couldn't say for sure what it was based on my description.[4] I wasn't surprised, though, because my only evidence was a half-century-old memory of a nondescript point of light observed by a young boy with no special knowledge on the subject.

Had I known what to look for in my backyard that night, and had I recorded everything—if not with a high-quality video camera, at least with pencil and paper—then I might be able to guess more confidently today, or at least to have given Mick West more to work with. But I can't even pin down the date or the time of day. Probably late June, probably eleven at night, but both may be way off. I also might have more accurately noted the thing's behavior. What direction was it? How many degrees across the sky did it traverse, and how high was it above the horizon?

What constellations were nearby? Was it constant in brightness? Did it change colors? How long between reversals of direction?

The *size-distance problem* is a major sticking point, too. Given an unfamiliar object, you can't judge its distance without knowing its size, *and vice versa*. But the human brain resists accepting such ambivalence. It tries to infer both size and distance by making assumptions based on visual cues, or memories, or expectations. Those inferences can be wickedly misleading.

What I saw may have been multiple lights close together at a great distance—members of a flying club moving in formation, for example. Or there may have been one small, dim light very close by. Mick suggested a Chinese lantern wafting in breezes.

I have two different explanations that I think are more plausible. First option: an advertising blimp with lighted signs. There used to be a lot more of them flying day and night, with many known cases of them being reported to police as UFOs. Far enough away, late enough in the evening, a blimp's sign is reduced to a single point of light against a dark sky. A typical flight pattern would take it back and forth over a crowd—a horse-racing track, for example, or other sporting event.

The second possibility is an airplane. Here's where recalling one small detail makes a difference: I remember the object's general direction. Looking at a map and considering my house's location and orientation, the UFO was due east, and I lived twenty-one miles due west of Logan International Airport. The light may have been a plane flying somewhere between me and Boston, shining its landing lights. These lights are white, non-blinking, and visible for up to around twenty miles on a clear night. They're sometimes used for extra visibility and safety in busy airspaces.

If it were a plane, why would it go back and forth? Search planes do so but not typically at night. More likely it was circling, and the back-and-forth movement was an optical illusion. It was too far away for us to see the standard red-and-green blinking wingtip lights but close enough that fore- and aft-facing landing lights were always in view.

There may be other possibilities. These are just what I can come up with today based on my recollections, my research on UFOs, and my minimal knowledge of aviation. I can't prove or disprove it was a blimp or a plane. Neither can I prove nor disprove that the pilot was an extraterrestrial being or "ET." If I wanted to believe in the ET hypothesis, nobody could prove me wrong. That doesn't make it any more likely to be right. Plus, given I have an alternative explanation that doesn't introduce assumptions about extraordinary beings or technologies, the ET hypothesis is very unlikely.[5]

One thing is for certain: I've treasured the indelible memory from my first camp-out experience. And all it cost me was having to straighten up my room.

UFOs definitely exist. This is true by definition. But we should keep in mind that the vast majority of them are either fully or plausibly identified by experts whose reports are far less popular than the tabloid stories and clickbait thumbnails screaming "*UFO!*" with their fuzzy or doctored photos. Still, there's always going to be a small subset of aerial phenomena that nobody can identify and for the same reason as my 1968 sighting: insufficient data.

To conclude that UFOs are evidence of intelligent ETs is a very different matter involving a massively illogical leap. It's saying, "I don't know what that thing is, therefore it's an alien spacecraft piloted by intelligent beings." Scientifically speaking, that's simply a bridge too far.

Better evidence than that is needed for the ET conclusion. A crashed saucer would be great—though not so much for its occupants. Or a message from space, such as the one realistically portrayed in the film *Contact*.[6] While many claim to have received such messages or witnessed crashed saucers, there's no reliable evidence of any of these types of encounters.

I'm not saying it can't happen. Some of my friends are surprised when I tell them I'd bet my life there are intelligent beings elsewhere in the known universe. Astronomy informs us that there are upward of two trillion galaxies, each averaging about a hundred billion stars. According to NASA, about half of those stars have planets. These solar systems offer a staggering array of natural laboratories for conducting ten billion years' worth of experiments on creating life-forms.

Time and space so far have limited our abilities to discover the results of any of those experiments—except one. Our own solar system is one-for-eight in terms of spawning intelligent life on one of its planets. If that's typical, then multiplying out the numbers in the last paragraph suggests there may be over a sextillion planets with intelligent life. The vast distances between us are what have so far stifled communication.

Don't expect those alien life-forms to look anything like us. Evolution by natural selection ensures this. Science fiction and alleged alien encounters are sadly lacking in imagination when portraying ETs as slightly modified humans with oversized eyes, swollen heads, and spindly limbs.

Even if intelligent alien life is highly likely, whether little green men (or women, genderfluids, or quadrisexuals) have stopped by Earth to have a peek at us is an entirely different issue. There's no decent evidence that intelligent aliens have ever visited. If UFOs are shy, ETs suffer from debilitating social anxiety.

Michael Shermer devotes his professional life to publishing and service on behalf of scientific skepticism.[7] Among his many projects is *The Skeptic Encyclopedia of Pseudoscience*. While developing it, he invited me to write the entry on UFOs.[8] He must have known that I was no expert and that there were more obvious choices for the piece.[9,10,11] What he did know was that, unlike most skeptical writing about UFOs, mine would lean more heavily on human factors: who or what leads people to believe in alien-piloted space vehicles despite the poor evidence.

It's not all about the brain, but a crucial factor is that we instinctively abhor a knowledge vacuum. Ambiguity unsettles us. The mind's desire to identify the unidentified is so powerful that it would rather settle for "mistaken" than suffer from "uncertain." Our conscious judgments and beliefs then rationalize our potentially misleading gut feelings. Often, to our own detriment, the brain doesn't care if it's wrong, as long as it feels right.

Social influences take over where the drive to quell uncertainty leaves off. We're not born with beliefs about vague objects in the sky. We learn from others how to interpret them. UFO buzz has been "a thing" in modern culture for seventy-five years. It's unavoidable. Currently, about 40 percent of Americans believe at least some UFOs are alien spacecraft, and that number is rising.[12] For various cultural reasons, beliefs vary across countries. For instance, only 22 percent of the British believe ETs have visited Earth,[13] while 47 percent of Canadians believe they've probably or definitely visited.[14] Nevertheless, they seem to be everywhere. Either there's something very human about leaping to the ET hypothesis, or the buggers have been coyly invading us all around the world.

Knowing what we know about people, human-centered explanations are the most plausible. Start with our sense of vision. As good as it is for detecting contrasts, movements, colors, and contours, its limitations are many. Misperceptions are most likely when objects are unfamiliar and observed under less-than-ideal conditions—true for most UFO sightings. Venus has been reported by car drivers as a UFO following alongside them as they travel down the road. The eye's natural movements can make a point of light appear to dart around against a dark sky. Optical illusions are an inevitable consequence of our sensory wiring.

A central aim of social psychology is to understand the many ways people influence each other. Especially interesting are situations where a subject adopts the beliefs of others without having any direct experience or evidence relevant to that belief. This effect is amplified when there are multiple others potentially influencing the subject, when they're physically and socially close to the subject, when the subject is emotionally aroused, and when the influencers have higher status or authority.

Recalling that fateful evening of my sighting with Jeff, the fact that he got a little amped-up when he saw the UFO was probably enough to get me excited too. Emotions are contagious. It's easy to find recordings of UFO sightings on YouTube where what's seen in the sky is no more spectacular than an airplane's lights but the observers can be heard whipping up one another's excitement levels. While emotional arousal makes an event more memorable, it also has some unfortunate consequences: It leads us to believe what happened was "real" even if it wasn't, and it distorts the perceptions, processing, and memory of what we think we saw.

Organizations can also play a critical role in legitimizing popular beliefs. One of the most relevant organizations is the US Air Force. Their job is to figure out whether UFOs are national security risks. They've published several reports over the years, always acknowledging that they can't identify all UFOs. Not all have sufficient data. But one person's lack of data is another's argument that the ET leap is warranted.

Recently, after decades without any government reports, the Office of the Director of National Intelligence released an eight-pager titled "Preliminary Assessment: Unidentified Aerial Phenomena."[15] Prior to its release in 2021, there was a great deal of hype in the media suggesting that, finally, we have evidence of spacecraft with advanced technology that are not of this planet. When it was finally released, the report landed with a thud. Completely boring. The line best capturing the findings was: ". . . we currently lack sufficient information in our dataset to attribute incidents to specific explanations. [*yawn*]"

No aliens.

Predictably, reports such as these trigger conspiracy theories about what the government *isn't* telling us. This further stokes public interest in UFOs, which then manifests in the form of local, regional, and national amateur UFO interest groups.[16]

One of these groups, the Mutual UFO Network (MUFON), emerged in the late 1960s in response to the burgeoning popularity of ET-related lore. It started small, of course, but now boasts chapters in every US state and forty-six countries.[17] It has an extensive database of cases, and anyone is free to report sightings, encounters, or abductions. Whereas a half-century ago MUFON's conception and birth were caused by a growing cultural phenomenon, today it helps to fuel and expand that very subculture, often with unfiltered, unskeptical information.

UFO proponents, such as those who join amateur investigation groups, commonly take a "draw the target around the bullseye" approach. That means starting with the conclusion: The unknown object is of alien origin. This leads the amateur investigator to ignore falsifying information and to overweigh anything that might support this belief such as rumors, folklore, hoaxes, fake documentaries, misperceptions, misuse of photo and video enhancements, and coincidences such as power outages or missing persons. They don't use evidence the way scientists do.[18]

Recently I reread my encyclopedia entry on UFOs for the first time in probably a decade. I have the same conclusions now as I did then, so I might as well just quote it:

> *The extraterrestrial hypothesis has found a stable niche in the ecology of public awareness. It is "locked in" in the sense that there is a critical mass of believers and promoters, sufficient to recruit new adherents and to sustain interest over time.... Because [... nonscientists] do not feel a need for any higher standards of evidence, UFOs are likely to persist as a cultural phenomenon even if proponents can amass no better evidence than that existing today.*

Most paranormal experiences are illusory. But even knowing this, we can see them for what they really are and treasure them just the same. The sense of wonder that Jeff and I experienced that night during our twelfth orbit of the sun was real. But it didn't make me leap to any baseless conclusions about extraterrestrials. Instead, it fed my desire to know more about the sky, the Earth, and life.

CHAPTER SIX

All Rise

The moment you doubt whether you can fly, you cease forever to be able to do it.

—J. M. Barrie[1]

FRONTSTAGE

Thirteen-year-old suburbanites have a low bar for excitement. Topping our list of "fun" activities are the parties. Usually parentally sanctioned, they're invariably held in underground rumpus rooms. We've seen more dimly lit basements than a true crime anthology series.

The parties are stupid, and I hate them. But the popular kids go, and girls go, so yeah, I go.

Party-talk surrounding each event is relentless. In the days before, we say, "You going to Greg's party? His parents are away. Gonna be so cool."

During, we tell each other, "Isn't this *the best*? Sucks for Greg he has to clean up before his parents get home. The beer was soooo good, right?"

And for a while after, it only improves with age. "Man, why weren't you at Greg's party Saturday? It was wild! His parents found out, and he's grounded for, like, a *year*."

There's no sex, drugs, or hard liquor at our "wild" parties, but there are other predictable indulgences. Much attention is paid to the vinyl 45s spinning on the portable record player. Great new music by Joplin, Otis Redding, Simon & Garfunkel, The Doors, Aretha, The Beatles. The volume's turned up all the way, pushing the player's six-inch speaker beyond its capacity and into realms of glorious distortion.

Greg's older sister has been deputized by his parents to keep an eye on things. Periodically, she peeks into the basement to ensure we're not destroying their home. Still too young and dumb for sex, we engage in sex-adjacent activities like the Twister game and slow dancing. I like these. I like being able to touch the girls without having to make small talk with them.

Drugs may not be a factor in these parties, but altered states are nevertheless achieved with gallons of sugary drinks and platefuls of sweet goods baked earlier by Greg's sister. Her motives had less to do with improving the undercover party and more to do with extorting Greg's lawn-mowing savings for her own upcoming secret events.

Sometimes there are minor violations of the no-alcohol policy. In this case, three cans of beer were smuggled in under puffy down jackets. These are lovingly and equitably doled out as though eighteen-year-old single malt Scotch. Nobody gets a buzz off of their ounce of beer, but we all think we do.

Our real gateways to altered states are party games that flirt with the paranormal. They usually start with group hypnosis. Our guide is a boy, Karl, whose voice has mostly changed to where his cracks and octave-leaps aren't too jarring. We turn off the music, gather round, and let him ply us with descriptions of quiet, natural settings. His dulcet tones soon lull us into a state of relaxation. Then Karl takes us on a journey backward in time. We imagine ourselves as children, then toddlers, infants, and embryos. Finally, we're merely souls, inhabiting unfamiliar bodies in past lives. Afterward, some of the kids swear they felt and saw *real* places and things from long ago, as though they were actually there. I'm open to these possibilities, but hypnosis simply doesn't work on me. It never takes me anywhere except to sleep.

Next up is a voodoo ceremony. It's a little eerie. It feels dangerous, even though it's never worked at any of the other parties. We fashion a doll out of some fabric scraps, socks, strings, and rubber bands. We name it Miss Terry after our most disliked teacher. We lower the lights and sit cross-legged on the floor, forming a circle around Miss Terry. Only about half the party's twenty attendees participate—or "play the game," as we refer to it. Each of us gets a sharp object or two such as pushpins, sewing needles, or corkscrews. We go around the circle recounting our negative experiences with Miss Terry and why she must suffer as a consequence. One of the girls leads us in a chant:

> *May she feel pain.*
> *May she regret.*
> *The wrongs she's done.*
> *And pay her debt.*

We repeat this ten times, then one by one, we ceremoniously stab the doll with our pointy object and leave it dangling out of her. After a moment of silence, we bring up the lights and laugh about it. We feel like criminals and secretly hope that it didn't work.

Next is the Ouija board. Even before we get started, many of the kids are chattering excitedly.

"It really works!"

"I'll watch, but I won't do it."

"You don't know what you're getting into."

I'm on the sidelines as first a group of three, followed by another of four, place their fingertips on the planchette. They pose their questions to the spirits and wait to see whether any responses are spelled out to them on the board.

The first group was unsuccessful. The planchette moved, but only randomly. The three members kept accusing one another of moving the planchette on purpose, though each swore they didn't.

The second group was excited to learn that a ghost named "A-B-E-L" inhabits Greg's family's house and died in the kitchen in 1789. I point out to them that the house is in the same development as my own, which was built fifteen years ago. They ignore me, and the session is considered a roaring success.

The last paranormal game is the weirdest of all: "Light as a feather, stiff as a board." In our version, one kid sits in an armless chair and four others stand around them. Greg the Partymeister is among the standing, and he's running the show. He explains that they'll attempt to levitate Rochelle using only two fingers of each hand. First, he asks them to try to lift her up this way. They slide their fingers under her arms and legs and try lifting, but she's squirmy and giggly and doesn't budge off of the chair.

When they all settle down again, Greg directs Rochelle to close her eyes, sit up "stiff as a board," and remain perfectly still. He tells the other three lifters to follow his lead.

"Now everyone, place your ten fingertips on Rochelle's head." They all follow obediently. "And . . . focus."

"Seriously?" says Larry, rolling his eyes. "This is so lame."

"FOCUS!" Greg repeats forcefully. He waits a moment and then, following Greg's lead, the lifters slowly circle Rochelle. They pause, keeping their fingertips on her head, while Greg slowly incants some magic words: "*Rizzenteel Upptalak.*"

They repeat this twice more—a slow orbit around Rochelle, pause, "*Rizzenteel Upptalak.*" Slow orbit, pause, magic words.

Now solemn and stationary, Greg shows them how to alternate their flattened palms in a stack above Rochelle's head, leaving an inch or so between each hand.

Greg looks up at the ceiling and instructs the spirits instead of the other kids.

"Spirits, *Rizzenteel* . . . make her buoyant! Spirits, *Upptalak* . . . take her weight! Spirits, *Rizzenteel-Upptalak* . . . may she rise into the air!"

Greg gives the others a nod. They place their fingers under Rochelle's legs and armpits and all together chant:

> *Light as a feather, stiff as a board.*
> *Light as a feather, stiff as a board.*
> *Light as a feather, stiff as a board.*

Finally, Greg makes eye contact with each lifter in turn and issues the order: "Levitate . . . NOW!"

Still in her seated position, Rochelle quickly rises a good two feet into the air, at which point some of the others' nervous laughter causes her to list sideways. They quickly lower her back down to her seat. Having been unable to budge her just moments before, the lifters are as amazed and enthralled as the onlookers. We'll be talking about this for days to come.

During the last phase of the party, we're all aglow with an "anything can happen" attitude. Greg's sister long ago tired of monitoring us. Now, spin the bottle and other kissing games are added to the party agenda. I'm inexperienced and nervous. But as required, bottles are spun, little heterosexual smooches are exchanged, and two by two the cooler kids sneak off to darkened corners to "suck face," as they'll refer to their make-out sessions when gloating about them back at school.

The six or seven of us who didn't pair up go back and huddle around the record player again. We take turns putting on songs that *really matter* to us. Then we dissect them, homing in on passages we love by guesstimating their locations in the tracks and manually dropping the tonearm's business end onto the spinning disc with an ear-ripping scratch.

The party winds down. As nine-thirty rolls around, those who live in the neighborhood say their goodbyes and walk home. The rest call their parents for rides and go wait for them at the curb along the sidewalk outside the house.

I walk home under the stars, lonely. I kick myself for once again not kissing any girls. But it's the first time I really listened to Ray Charles and Janis Ian records, and now their songs are going through my head. It wasn't a total loss.

Backstage

Susane Colasanti describes the agony of adolescence this way:

> *Standing in line at the food court, I try to be myself. But I forget how I usually stand when I'm myself.*[2]

Lindsey Leavitt sums it all up:

> *Adolescence is the same tragedy being performed again and again. The only things that change are the stage props.*[3]

It's a time of massive physical, emotional, and cognitive transformation, crammed into a very small slice of the human life. This upheaval isn't for the faint of heart, and few emerge unscathed. I can attest that my stress and anxiety levels at times reached heights comparable to the very worst experienced in my subsequent adult life. There is a bottomlessness to adolescent lows, an unboundedness to the highs, and adults forget how little it takes to trigger either extreme.

Susane Colasanti pinpointed the more specific angst of feeling uneasy in one's own body and being unsure about basic aspects of one's identity. Again, drawing on personal experience, I can remember being at multiple personality and behavioral crossroads where I paused to ask myself: *What kind of person do I want to be? How do I want others to see me?* And with so little developmental momentum behind me, choices I made at those crossroads actually made a difference. They changed me.

In one case, I'd been hanging out with several boys with whom I cruelly targeted and "ranked-out" a few perfectly innocent victims. I derived some pleasure from this bullying. Feelings of superiority and camaraderie. But eventually, I decided I didn't like doing it anymore and, literally overnight, quit the gang and successfully redefined myself as a nicer person. What I did before was inexcusable, and to this day, it is one of my very few regrets in life.

Most of us grow up in religions that teach us stories of miracles and reality-defying supernatural beings. Depending on the family you were born into, these may include gods, angels, devils, demons, djinns, kachinas, zombies, witches, or others. But it's a whole other thing when a good friend looks you in the eye and claims to have seen a real ghost, or a girl you like is into astrology, asks "What's your sign?", and believes she then knows fundamental truths about your personality. Our first encounters with peers espousing nonreligious paranormal claims often coincide with the times our identities are malleable and our social lives are fraught with quicksand.

Several factors combine to make teens more vulnerable to peer influences than children or adults. For example, research on risk-taking finds that *homophily* and *peer socialization* have major effects.[4] Homophily (pronounced *ha-MOFF-a-lee*) is the tendency to associate with others similar to yourself. So if you're interested in Ouija boards, you're more likely to hang with others who'll reinforce this interest. "Socialization" means learning a group's norms for acceptable behavior. If your peer group happens to be interested in Ouija boards, then you'll tend to acquire the group's attitudes and beliefs about communicating with spirits.

I came to the parties for the chance to feel normal and kiss a girl, but instead, I was exposed to past lives, spirits, voodoo torture, and levitation. Let's take a quick look at each extraordinary claim.

Hypnosis. Psychologists are deeply divided about the nature of hypnosis. After decades of debate, they haven't even settled whether or not it's an "altered state of consciousness,"[5] or how to test objectively whether a person is under hypnosis. But most agree with statements by the American Psychological Association that hypnosis makes some people more suggestible.[6] A side benefit is that it can help gastrointestinal disorders, skin conditions, anxiety, phobias, and chronic pain.

Very few psychologists believe that hypnosis works for past-life meet and greets. There are lots of personal testimonials from people who are convinced their souls once inhabited other bodies, but even when testimonials seem compelling and sincere, they don't warrant our belief. Not only does the claim require a series of unlikely assumptions to be true, but even the best scientific research supporting reincarnation fails to survive careful scrutiny.[7,8]

What is known for certain is that hypnosis enables imaginative storytelling and false memories.[9] Even when events transpiring under hypnosis seem real to the experiencer, they're explainable as psychological rather than paranormal. These facts make it no less exhilarating—or relaxing, as the case may be—when seventh graders play with hypnosis at a party. They also may find it more comforting to see themselves as the heroes in confabulations of past lives than to confront the nerve-wracking realities of adolescence.

Voodoo. Voodoo dolls were portrayed as powerful tools of evil in the 1984 film *Indiana Jones and the Temple of Doom*. Indy appears to be losing a fight while on a moving conveyor belt heading into a noisy rock crusher. Nearby, a sort of priestess character incapacitates him by stabbing a stiletto into a voodoo effigy. When his rescuer liberates the doll and removes its dagger, Indy bolts upright with renewed strength and proceeds to save himself.

About the only similarity between Indiana Jones's professional life and mine is that we've collected some artifacts relating to our work. Mine are little gifts from students who took my Sociology of the Paranormal course. One toy looks like a foggy snow globe, but when you wind it up, an alien pops out of an egg like a clown out of a jack-in-the-box. I was also given a sort of cookie-cutter thing that you press lightly into a slice of bread. Put it in the toaster and the slice emerges with the outline of the Virgin Mary etched on one side.

One of the most striking knickknacks is a soft, bright red, faceless doll that came with multiple common pins stuck into it. I actually never taught about voodoo dolls, or the Vodou tradition in Haiti, or the Voodoo religion still practiced in the American South.[10] Such a gift informs me that voodoo dolls are still a thing in our collective psyche. My intuition is bolstered further by opinion polls showing that about 20 percent of Americans believe in some form of witchcraft—a generic term that sometimes encompasses voodoo practice.[11]

Although the Indiana Jones franchise arrived years after my adolescence, it turns out that voodoo was not an uncommon feature in television and films during the years leading up to my middle-school partying. Neither was it really mainstream. For kids coming of age in the late 1960s, we were tantalized by its aura of power and danger.

In reality, there is little connection between cultural Voodoo practices and the use of voodoo dolls.[12] Nor is there any evidence that the dolls work as advertised. It's possible that voodoo curses could induce a self-fulfilling prophecy if one who knows they've been targeted, believes in such things, and becomes fearful to the point of self-sabotage. A post by a Quora.com user offers a simpler way to potentially harm someone with a voodoo doll: Throw it at the intended target with sufficient force.[13]

Ouija Boards. If you have any doubts about your Ouija board's ability to bridge "The Great Divide" between the living and the dead, a thousand videos on the Internet show them in action and attest to their robustness. There you can find giddy spirit seekers calling on dead people who never tire of answering excruciatingly mundane questions such as "Are you there?"

Ouija boards haven't changed much in the last 130 years. About the size of a standard board game, on one side are printed the digits zero through nine, the alphabet, and the words "YES," "NO," and "GOOD BYE." It comes with a hand-sized planchette with felt or caster feet for easy gliding around the board. Two or more users work the planchette by placing their fingertips on its edges, asking questions of the spirits and reading their replies as a series of letters or numbers (or YESes and NOs) indicated by the planchette during pauses in its movements.

I've never had a Ouija board give discernible answers to my questions. When answers to others' questions arrived on cue, it always appeared to me that they were purposely moving the planchette. But that might not have been true. As it turns out, Ouija boards can appear to work because of unconscious hand movements—the so-called *ideomotor effect* by which subtle physical actions may occur in accord with suggestions or unconscious expectations.[14] If users are answering their own questions with unconscious hand movements, it should be easy to demonstrate this. Why not blindfold them and simply let the spirits guide the planchette?

Mentalist and debunker Mark Edward did just that on an episode of the National Geographic Channel's *Brain Games*.[15] The spirits answered basic questions when the four users could see the responses. As soon as they were blindfolded, the planchette was no less active, but it kept landing between numbers or on blank areas. The spirits didn't give a single clear answer to the same questions that they answered so easily when the participants could see what the planchette was doing.

In a similar vein, mentalist Derren Brown showed that, when people in an allegedly haunted house were led to believe a totally contrived ghost story, they generated Ouija responses exactly in accord with that story. And still, none of the subjects believed they were moving the planchette.[16]

That leaves three possibilities: (1) At least one subject was lying; (2) at least one moved the planchette unconsciously; or (3) spirits moved the planchette after popping into existence to play out roles in a fictitious story. We can't prove that the third possibility is wrong. To accept it, however, would suggest a deep desire to shun the simple, demonstrable, non-paranormal explanation in favor of the paranormal one that lacks any evidence.

I've heard that spirits don't like skeptics and prefer not to reveal themselves when a skeptic is present. Maybe so, but what would they be afraid of? It's not as though my questions could possibly inflict any harm. I'd just *beg* them to explain their nature to us so that we could understand and appreciate them better. Is that too much to ask? "YES," apparently.

Light as a Feather, Stiff as a Board. Hailey Reese has over a million subscribers to her paranormal-themed YouTube channel. The thumbnail image for her episode on the light-as-a-feather party game screams "DO NOT PLAY" in bold red letters. She incorporates a series of disclaimers and warnings about the dangers:

> *I want to say that when you are opening doors to the unknown, it is exactly that. The unknown. And it's not always as easy as hoping for something positive to occur. So just really really be cautious when you are playing paranormal games if it is something that you decide to do. Protect yourself, make sure that you've read the rules properly, and just be as safe as possible.*[17]

The biggest danger in toying with paranormal games like light as a feather, she explains, is that the levitated person must open their body to being possessed by spirits that "may not want to leave." This would be perfectly sensible *if* spirit possession were the best explanation for levitation at teen parties and sleepovers. It's not.

Let's start off with a look at several kinds of levitation claims floating around. Then we'll return to light as a feather.

The Balducci Technique is one of the simplest levitation tricks. I even pulled it off at the front of a classroom full of students. It was during a unit on perception, and I made it seem impromptu. In fact, I'd done a little advance preparation with my teaching assistant. He helped to ensure that when I stood in a certain spot, the front desk blocked my lower legs for most students but allowed those directly in front of me to see my feet.

In the middle of the session, I informed the students that I wanted to try something new that was *not* a trick of perception. I would raise my entire body about two inches off the floor. I had their attention.

You may want to try this yourself as you read the description. I got into position with my left side toward the students. Then I slowly pushed down against the ball of my right foot, lifting up my right heel—and entire body—by a couple of inches. (It's okay to be a little off-balance and wobble. You're floating in the air, after all!) Because my left foot blocked their view of the front of my right, all the class could see were my heels and my entire left foot leaving the floor. Students with obstructed views still saw both my body rise up and the expressions of the students who had a clearer view. Few, if any, could see how the levitation was accomplished. They responded with smiles, laughter, gasps, and scattered applause. We had a nice little discussion during which I revealed the technique. It drove home the point that something can appear to violate the law of gravity quite convincingly—if you don't know how it's done.

A second and more elaborate levitation illusion is the Hovering Monk. The effect is striking even when you know how it's done and absolutely stupefying if you don't. There are many variations and methods (easy to find by searching social media[18]), but most are accomplished using a purpose-built steel frame to do the heavy lifting. The upper portion of the frame hides beneath the floater's loose-fitting clothing. A vertical support is disguised by a cane or wooden stick. The base is concealed with a blanket or carpet on the ground.

Levitation tricks are also standard fare for stage magicians. David Copperfield performed several variations in a series of nationally televised specials between 1977 and 2001. His amazing visuals combined both levitating and soaring through the air—in one case, around a large stage,[19] and in another, above the Grand Canyon.[20] For Copperfield, television had two great advantages over the stage: it afforded total control over the viewers' perspective, and the lower-resolution broadcasts of the day masked crucial details about how the tricks were accomplished.

Because the inventor of Copperfield's flying apparatus was granted a US patent,[21] we know exactly how the flying trick was performed: harnesses, wires, gimbals, levers, and other mechanical components, all motion-controlled by computer. It's a clever piece of engineering that works with the laws of physics rather than against them. It distributes the flyer's weight amongst numerous fine wires, allowing them to virtually disappear against the background.[22] Like the Hovering Monk, the effect is no less stunning for knowing how it's done.

The explanation for the light-as-a-feather levitation is simplicity itself. No laws of physics are broken here either. The person being lifted doesn't experience a sudden weight loss. We can't rule out spirit possession, but that method seems unduly complicated and unreliable compared to what's really going on.

The lift fails initially because the liftee's body is relaxed and the lifters aren't coordinated. The ritual takes care of these problems. First, the liftee's stiffness ensures a more even weight distribution across the lifters' eight hands, with each hand needing to briefly raise only fifteen or twenty pounds. Second, the pre-lift ritual coordinates the lifters' efforts at just the crucial moment. And that's it. Balance and coordination.

Lifting a person out of a chair is not exactly what you'd call a mystical experience. It's the incantations, group chants, coordinated rituals, focused attention, and a sense of shared purpose, which combine to invoke a collective sense of otherworldliness. There's also likely to be an adrenaline kick further enhancing the lifters' strength and excitement.[23] The ritual transforms the situation from just a bunch of kids lifting up another kid into a flirtation with the mystical. Truly, it is an extraordinary experience.

Peer influence starts to take hold well before high school and is already in full swing when we navigate the strange land of junior high. There we discover unforeseen humiliations and inhabit awkward, acne-riddled bodies. We try to remember how to act normal and fit in and to figure out who we want to be—usually someone else. Racked with insecurity, we surrender to peer pressure's worst intentions.

We're gradually leaving parents behind as we take over the job of raising ourselves, and each other.

Yes, we raise each other, sharing and internalizing unspoken dress codes, hair requirements, complicated school-dance dynamics, and the dark web of clandestine notes passed behind teachers' backs.

We raise each other with unrequited crushes and first kisses.

And sometimes, we raise each other through levitation rituals at obligatory parties that connect us with the extraordinary power of each other.

CHAPTER SEVEN

Onward Christian Scientists

Man ... is the only animal that has the True Religion. Several of them.
—Mark Twain[1]

Frontstage
I'm a proud Roister Doister.

Back in high school, I had a very successful run in theatrical productions staged throughout my junior and senior years. Now, as a freshman at UMass/Amherst, I'm batting zero after multiple try-outs for Theater Department productions. Not unreasonably, they mainly cast their own students. Eventually, I find steady "work" with the Roister Doisters theater club. Not nearly so serious as the Theater Department, Roister Doisters lets me keep one foot on stage while pursuing my regular studies.

Our current production is an original musical, *Dickie the Dam and the Big Blue River*. It's ambitious. Tom Keegan, its author, lyricist, and composer, is a fellow UMass student and happens to be an alum from my high school. He graduated three years ahead of me, and we'd not met before this semester. But we have a number of mutual friends and acquaintances back home. He's a fellow guitarist and is always nice to me despite my status as a lowly freshman.

My best new friend since starting college, Richie, won the role of Dickie the Dam. Great actor, weak singer, but also great at acting like a singer. Last year, he played the lead in his community theater's production of *Godspell*. In our show, he's proving to be charming and convincing as he creates the role of an anthropomorphic dam.

I was cast in a decent supporting role as one of two "Tall Towers" flanking Dickie at all times. I've got a rousing duet with the other tower, quite a few spoken lines, some chorus numbers, and I get to play guitar in the orchestra pit on several songs.

This big, weird show may bomb, but working on it is a blast. For two weeks, we've rehearsed three hours every weekday evening and up to eight hours a day on weekends. With a month still yet to go, we've already become a family.

Wayne is my newest theater buddy, a seriously easy-going guy in a persistently good mood. At six-four, he's got two inches and at least fifty pounds on me. Despite his size, he moves with grace both on and off the stage. His positive vibe makes him an arresting figure to audiences. He's miscast and underutilized in the chorus where, through no fault of his own, he draws undue visual attention during group numbers. Not having a speaking role means he's often free to socialize during rehearsals. People gravitate to him like he's a star. Our senses of humor mesh, and we get along particularly well.

On a Friday a couple of weeks into rehearsals, he buttonholes me on the far side of our practice space. I've been hanging out on the periphery with a sloppy head cold. Wayne approaches.

"Hey, Barry! I was talking to my housemates about the show and the cool people in the cast. They'd like to meet you. You wanna come over for dinner tomorrow after *rehassle*?" That's our little inside joke word for the thirty hours a week we're in here. "We do group meals a couple times a week. It's fun."

"So you couldn't get any cool people to come, and you're asking *me* now?" He laughs at that. "Yeah, I'd really like to, actually, but obviously . . . this cold . . . I don't want to infect anyone."

He dismisses my prudence with a wave of his hand, insisting, "No, it's okay. You should come tomorrow if you're up to it. It'll be good for you."

"Will it work to wait and see how I am tomorrow?"

"No problem," he says cheerfully.

Next day, the cold's no worse, so I accept the invitation. Wayne jots down directions for me during a break, and eventually, I finish out the long day's work of pretending to be a jaunty civil engineering project.

Around six o'clock, I drive to Wayne's place on the edge of town. I spot the street number on the mailbox at the end of his driveway. Set back around a hundred feet from the road is a tidy little colonial-style home. It's a far cry from the usual "Animal House" dump where most off-campus students live. I park, approach the front door, and ring the bell. After a moment, the door swings open, and Wayne invites me in.

Standing behind him are five more young adults. They're all smiling, unnervingly, standing there like full-size cardboard cutouts masquerading as real boys and girls. Wayne introduces them collectively as his housemates, then individually by name. One by one, they come to life with a little wave or friendly salute. Becky, Ollie, Mike, Mindy, and Ted.

Adding to the feeling that something's slightly off, they continue standing there smiling, wordlessly, five seconds too long. Finally, they scatter like billiard balls in an opening break. "Dinner at six-thirty!" Wayne shouts before they're out of sight.

Wayne gives me a quick tour of the house. Upstairs are three nearly identical bedrooms, each with a bunk bed, two dressers, and two desks. There are a few personal items on surfaces—framed photos, ornaments, a few books and papers. No posters or artwork on the walls. It's impressively neat and austere. As elsewhere in the home, there are a few religious items on windowsills, walls, and tabletops. Nothing gaudy. Jesus and Mary figurines, a candleholder shaped like a bible, a small cross on an otherwise blank wall.

Down on the main floor, Mike, Ollie, and Ted are busily preparing food in the kitchen. Mindy and Becky set the table in the dining area. Wayne and I cross to the living room where a fire crackles in the hearth. He grabs a photo album from a low shelf, and we sit together on the couch. He wants to show me snapshots from a retreat he went on last summer.

"What kind of retreat was it?" I ask.

"My church group," he replies while turning a page.

"Really beautiful! Where was it?"

"Outside of Denver. These photos were in Rocky Mountain National Park."

"Incredible!" It truly is. "I've never been there. What church are you with?"

Wayne looked up at me and said, "I'm a Christian Scientist. So are my housemates."

"Oh, cool. I don't think I know any other Christian Scientists. I've visited that incredible complex in Boston—with the Christian Science Monitor and the library. Twice on school trips. And the 'Maparium,' where you walk inside the stained-glass sphere and look at a map of the globe from the inside out? Amazing!"

"Yeah, that's like our 'Mecca,'" Wayne says with a chuckle.

"There was also this Christian Science Reading Room in downtown Framingham. It was right by the place I worked for two years in high school. I walked by their window hundreds of times, but it always looked empty. I never went in."

"I'm not surprised," he remarks. "A lot of those places are closing. Our church is shrinking." Changing subjects, he asks, "How's your cold today?"

"It's hanging on but not too bad," I report.

"You know, we can definitely help you with that," he offers in a very upbeat tone.

I know what he's talking about, and it's not chicken soup. He's offering to coordinate with his housemates to faith-heal my cold. I know this because of what I learned on my field trips to Boston's Mecca of Christian Science: They reject conventional medicine and are fully committed to treating all illnesses with prayer.

What the hell, I probably should say yes. But I'm both wary and weary. This semester alone, I've been approached by zealous recruiters for so many religious groups—the Hare Krishnas, Jews for Jesus, Transcendental Meditation, Mormons, Campus Crusade for Christ, Hillel Society, Newman Center, and others. Might these friendly Christian Scientists also have designs on my eternal soul? Is their home cooking my apple of temptation?

"No, thanks. I'm good," I reply.

The call comes from the kitchen: "Dinner's on!"

The dining area between the kitchen and living room is crammed tight with a rustic farmhouse table and eight chairs. They motion me to one end, flanked by three housemates on either side. Being at the head makes me the center of attention, but not unpleasantly so. Even before everyone's seated, they're asking questions about my background and freshman experiences.

Wayne's tasked with serving. "Welsh Rabbit," he explains as he first distributes a slice of rye toast to each of the seven plates. I think he notices my expression and adds, "No rabbits were harmed in the making of this meal. Anyway, the real name is 'rare-bit,' not 'rabbit.'" He fetches a soup tureen, and each bread slice gets coated with an exact ladleful of the gooey concoction.

When everyone's served, they decide it's Wayne's turn to say grace. They all bow their heads while he utters a standard pre-meal prayer. I lower my head respectfully.

Oliver, the *chef du jour*, explains the recipe as we start to dig in. "It's mostly melted cheese and milk, with some Worcestershire sauce and mustard. Some people use beer. Not us." The others snicker at this. "The bread's fresh this morning from the bakery on North Main."

We eat, and we talk. The meal is simple and tasty. The housemates loosen up and ease into more natural behaviors. I don't think anyone's consciously avoiding the topic of religion, or Christian Science in particular, but neither do we pursue it in any substantial way.

Most days, I have three meals in the dining hall near my dorm. There's a different cast of characters at the table every time, quite unlike the dining experience here in Wayne's world. I'm enjoying the change of scene, watching this group's friendly and familiar dynamics. Six different personalities, like six guitar strings playing different tones, all harmonizing on the same instrument. They incorporate my occasional grace note, and it feels good to blend in with them.

Wayne clears the table and graciously announces that there's tea, coffee, or hot chocolate on the kitchen counter. As we're getting our hot beverages, Wayne returns with individual servings of ice cream on shortcake. After we've gobbled that down, I move to clear the table after we're all finished. I'm stopped by Mindy and Becky. Gesturing toward the duty roster hanging on the refrigerator nearby, they explain they're on dish duty and will take care of the cleanup.

The rest of the dinner group moves from table to fireside. The housemates are intensely curious about the *Dickie the Dam* project. Wayne and I answer their questions, supplemented with gossip about some of the burgeoning romances and backstage shenanigans.

Wayne gets up and grabs a guitar from a floor stand in the corner. Handing it to me, he suggests, "Let's do a couple songs from the show." I play the intro to the title song, and we start singing the long, descriptive ode to Dicky the Dam. By the last verse of our second song, only the two of us are left in the room. I use that as my cue to head home. I thank Wayne for inviting me, ask him to thank the others for their kindness, and say that I hope nobody catches my cold.

"No worries," he assures me. "We don't get sick."

Backstage

The top-rated sitcom in 1965 was *The Beverly Hillbillies*. A decade later, it was still in reruns. One episode keyed on Granny's vaunted cure for the common cold. The fish-out-of-water family's matriarch had forged a deal with a pharmaceutical sales rep to distribute her cold remedy up and down the West Coast. The rep took the family's word that the cure hadn't failed in forty-five years. But it all fell apart the moment Granny and son-in-law Jed gave directions for using the potion: "Take one spoonful of cold cure, eat sensible, get lots of rest, and drink plenty of water. In a week or ten days, your cold will be gone! That's all it takes."[2]

I'll call this "the Granny Effect."

The joke plays on the fact that Granny's treatment's effect is illusory. The body eventually cures itself of minor ailments like colds and also many more serious ones. But rather than credit the autoimmune system, sometimes it *feels* reasonable to attribute the cure to something else—a benign potion, or in the case of Christian Scientists, a personal, loving god. Whether you administer Granny's snake oil or prayer, your cold will be cured and your beliefs in the treatment's efficacy will appear to have been affirmed.

The Church of Christ, Scientist was established in 1879 by Mary Baker Eddy. She laid out her rationale in her book *Science and Health with Key to the Scriptures*. She claims that Christian Science effects cures through the mind, citing the "divine Mind" as the wellspring of power. Prayer heals, Mrs. Eddy insisted, by replacing your faulty beliefs with her correct beliefs. By many accounts, she was a charismatic bully, intolerant of anyone who contradicted her. She declared herself to be the supreme healer, infallible as Christ, and many believed her.[3] For a period in the 1930s, The Church of Christ, Scientist was the fastest-growing religion in the United States.[4]

Eddy claimed that every follower could be taught to heal the sick and that illness is only an illusion stemming from insufficient faith.[5] Forget about colds. Try to imagine the magnitude of faith required to accept that your gangrenous limb, crippling arthritis, skin cancer, brain tumor, or STD is only an illusion—and *always* your fault for not having enough faith.

In case it needs saying, Christian Science is no more a science than Buffalo, New York, is a buffalo. That's not to say it can't work. It's just weird that "science" has been incorporated into the name.

Christian Scientists contend that modern medicine is inferior to prayer and physicians promote disease by being ignorant of the fact that the human mind and body are only myths. The church opposes medicines, vaccinations, and quarantines. They reject any means of relieving discomfort and pain—even back rubs or ice packs.[6] That said, they do sometimes make exceptions that, to me, sound like doublespeak. For example, the church seeks to exempt members from mandated vaccinations but also claims to support the goals of public health.

There is no evidence supporting the claim that Christian Science healing works better than standard medical treatments. When Christian Science healing *seems* to work, it's a combination of the Granny Effect—the autoimmune system doing its job—and the placebo effect. And when praying seems *not* to work, socially encouraged victim blaming takes up the slack. Christian Scientists can rationalize away failures and cling to the belief that "for over a century, thousands of families and individuals have confidently, safely, and successfully relied on Christian Science for healing of every sort."[7]

No healthcare system is perfect. But research has yielded proven remedies for many conditions that become dangerous if left untreated:

diabetes, meningitis, peritonitis, and ruptured appendix, to name a few. Sometimes a person—often a child—needlessly suffers or dies because prayer was applied rather than proven treatments.[8,9,10] Unfortunately, Christian Scientists live shorter lives as a consequence of their beliefs.[11]

A dying membership is only one of the factors contributing to the church's decline over the last fifty years. Others include members renouncing church practices as medical science advances; weak socialization of church youth; financial issues at the organizational level; and the severity of church doctrines.[12] The number of practitioners and teachers in the United States is down over 80 percent, with a 60 percent decline in the number of churches.[13] In the United States, it is likely that fewer than forty thousand church members remain.[14]

Christian Scientists are relatively extreme in their views, but a great majority of Americans—and countless millions of others around the world—believe in some form of spiritual healing. Some versions call upon one or more gods to do the job. Others summon mysterious healing "energy" from within the person, or from someone or something in the environment. Many practitioners are fine with working alongside conventional medical treatment. Others discourage their clients from seeking it at all.

Praying comforts believers and makes them feel good. I'm a believer in the healing power of feeling good. The sense of comfort and the social support in religious contexts are very likely to have health benefits, both physical and mental.[15,16] There's also a growing body of research studying direct physiological connections between brain activity and autoimmune system functioning. In this way, for example, the brain's ability to regulate emotion impacts inflammation in the body.[17]

More extreme is the assumption that one person's prayers can heal other people remotely. If believers are willing to pray rather than allow themselves or their children to take proven remedies, praying would appear to become a potentially deadly practice. What does the evidence show?

A great majority of adults in the United States pray to a god, most on a daily basis.[18] They pray for different reasons: to give thanks, ask for personal help, pay homage, confess sins, or help another in need. And of course, people pray for desired things to happen. Sometimes I even catch myself pleading to a higher power to help me find my missing wallet and thanking it when I locate the wallet between my couch cushions.

As a side note, given there are so many different beliefs about supreme beings, I'll use "god" or "it" to refer to generic deities, and "God" or "Him" to indicate the Judeo-Christian god.

The idea of praying for something is simple enough. You speak to your god in your native tongue, typically in your private thoughts or quietly aloud. You may do this alone or with others. You ask your god to do something that, without its intervention, you assume would be less likely to occur. Examples might include asking for an end to homelessness, a cure for your friend's cancer, a win for your sports team, an enemy's slow and painful death, or a million-dollar windfall. Although praying may seem petty when it's for a sports team, or immoral when for someone's death, or selfish when for sudden wealth, that's really for the prayed-to god to judge.

These are all so-called intercessory prayers. What supposedly happens is that a person or group prays for x; the god receives and responds to the prayer; and the god makes x happen. It's the simplest form of cause-and-effect claim. Sure, one of the links is supernatural, but the logic is simple enough: person prays for x, then the god receives the prayer and makes x happen.

I see a claim like this and all I can think is "You can test this!" And not surprisingly, dozens of science-minded believers have done so. Perhaps they wanted to prove to skeptics that praying works or use their findings to evangelize. Or maybe they simply wanted to validate their own faith. The motivation won't matter if the research is designed properly.

When teaching about research methods, I've sometimes used published studies on intercessory praying to demonstrate both good and bad research practices. Studies that claim success typically have either very small effects or very big design flaws. An example of a design flaw is having subjects who are sick be aware they're being prayed for. This knowledge may give comfort and benefit them independent of any divine intervention or lead them to overstate their improvement.

Another problem that plagues many prayer studies is having too few subjects. This makes it more likely that the occasional positive result is due to chance rather than prayer. Another practice that I've seen is when researchers measure dozens of different factors but cherry-pick just the ones that moved in a direction favoring their beliefs. Results that drift the other way tend not to be reported.

In his extensive search for published research on intercessory prayer, David Hodge found just seventeen studies that were both relevant and minimally well designed.[19] Some of those showed positive effects of prayer,[20] but the majority didn't. One study in particular by Herbert Benson and associates was exemplary, the most scientifically rigorous by far.[21] I'll sketch it here briefly. Notably, it was sponsored by a religious foundation, and Benson himself is a believer. But the research team took precautions to prevent religious biases from influencing their results.

Prayers were offered by members of three Christian prayer groups. About eighteen hundred cardiac bypass patients from six hospitals were assigned randomly to one of three target groups. Members of target Groups 1 and 2 received intercessory prayers. Group 3, the control group, did not. People in Groups 1 and 3 didn't know whether or not they were being prayed for; Group 2 members were informed they were prayed for. Here's the design:

	Group 1	**Group 2**	**Group 3**
prayed for	Yes	Yes	No
knew it	No	Yes	No

If prayers work, it shouldn't matter whether or not recipients know they're being prayed for. So both Groups 1 and 2 should fare better than Group 3. But that's not what was found. There was no difference between Groups 1 and 3. Cardiac patients received no benefit from prayers compared to those who weren't prayed for. Strangely enough, Group 2 members—who were prayed for and knew it—had significantly *worse* complications than Group 3. Maybe Group 2 members felt some performance anxiety from knowing they were prayed for and not wanting to fail those who were trying to help them.

If not a supreme being, there's at least a supreme irony here. If the Christian God holds the reins of all earthly endeavors, then why did He cause Benson's beautifully constructed experiment to show that praying does no good? God missed an opportunity to make a powerful statement: allowing scientists to demonstrate the power of prayer, or better still, miraculously curing eighteen hundred ailing patients across the three conditions. The usual excuse is that God is testing believers' faith—an excuse that the researchers *did not* try to make. Doing so would have assumed knowledge of God's mind without evidence and committed the *ad hoc fallacy* by making up a just-so story after the fact. Plus, I find it hard to swallow that a loving God would sacrifice the health of twelve hundred patients in Groups 1 and 2 just to perform His own experimental test on the faithful experimenters. The simplest interpretation remains that the prayers didn't work.

If there's a god, I wouldn't be so arrogant as to claim I understand its thought processes. So I wonder by what authority some—like Mary Baker Eddy—claim to know the unknowable?

For sociologists, that's not a rhetorical question. According to *legitimacy theory*, a figure's authority is the product of group consensus.[22] The consensus may emerge when members see it as *proper* or *expeditious* to grant someone authority over them. So a preacher, president, or god becomes legitimate when group members either think it's right that they're in their position, or when they think it's easier to go along with the situation than to buck authority and risk social costs. In no case does legitimacy theory presume that one whose decrees or actions hold sway over others has a divine right to do so.

The implication is that since authority is socially constructed, it can be socially demolished when undeserved. We needn't believe in the power of prayer just because authorities have always told us to do so, or because it's easier not to go against the grain. But don't assume I'm an authority, either. Use your God-given or evolution-given brain and decide for yourself.

Although I declined the Christian Scientists' prayerful intervention in the predictable arc of my cold, I was humbled by the few hours I spent with them. They joyfully broke bread with me. They treated me, a stranger, with warmth, kindness, and respect. The memory of that evening has lasted decades.

How to square this with the misery inflicted by a belief system which denies that God sent us brilliant medical researchers and practitioners to treat very real illnesses and traumas?

Many readers have probably heard the story of "Two Boats and a Helicopter."[23] A preacher is caught in a storm at his church. Floodwaters rise, and he prays for God to save him. When someone in a rowboat offers assistance, the preacher declines saying he's waiting for God's help. As the waters rise further, he turns down another good Samaritan in a motorboat. Then, with only the steeple above water, he clings for dear life. A pilot in a helicopter swoops down to help, but again, the preacher refuses. Soon thereafter, he succumbs to the floodwaters and drowns. When he gets to heaven, he asks God, "Why didn't you save me from the flood?" To which God replies, "What more did you want from me? I sent you two boats and a helicopter!"

Members of the Church of Christ, Scientist could have chosen to believe that God delivers relief from pain and suffering via doctors and medicine. Mary Baker Eddy told them otherwise, and they accepted her authority.

At the Tuesday rehearsal after the Saturday dinner, Wayne's voice was noticeably raspy. By Wednesday, it was obvious he was coming down with a cold. He told me four out of five of his housemates were in the same condition. "Maybe some sort of allergy," he suggested.

I said, "Wayne, I'm pretty sure you all caught my cold. I am so sorry for infecting everyone!"

He would have none of that. "If we're actually sick, it wasn't your fault," he said. "It just means we weren't disciplined enough."

I wished him a speedy recovery and asked him to tell his housemates the same.

I had thoughts of dismissing Wayne and his housemates as acting foolish, oblivious to reality, blinded by an absurd set of beliefs. But that wasn't so easy. They weren't fools, and they were sweet and kind. If everyone were more like them, our society would be more peaceful and considerate. Except there'd be all that needless suffering. I don't know how to reconcile this dichotomy beyond trying to shed light on it. I remind myself that you don't have to hold untenable beliefs in order to treat people with warmth and respect. Anyone can do it.

Also, I would suggest that others use their God-given critical thinking abilities when making medical decisions—at least for their kids' sake.

Finally, not all authorities are created equal, so be wary of those who are just making shit up.

I wonder if Wayne and his nice housemates ever allowed for the possibility that I gave them the cold virus. Probably not. If true to their faith, the only things I might have infected them with were bad thoughts.

Undoubtedly, they prayed, and on the seventh day or thereabouts, God cured them.

CHAPTER EIGHT

He's Pulling Her Leg

I went to a faith healing show last night. It was so bad, even the guy in the wheelchair walked out.

—Unknown

Frontstage
Rehearsals are underway for my high school's production of *Hello, Dolly!* I'm playing Horace Vandergelder, the show's male lead and my first big role. My acting career began only a few months ago with the goofy little melodrama *Curses, the Villain Is Foiled*. I was Godfrey Goodheart, Tess Truly's love interest. The only reason I started doing this was in hopes that acting like someone else would help me overcome a deterrent fear of speaking in front of people. Hiding behind theatrical masks worked for me and literally changed my life.

Curses had a cast of six. *Dolly*'s is twice the size and a much bigger production with its dancers, chorus, and orchestra. I'm overwhelmed by how much there is to learn. I keep my head down and work hard every day on my lines, songs, steps, positions, and cues. I try not to think about the public humiliation I could face in eight weeks.

Emily is in the show, too. A smart and pretty sophomore, she's shyer than the more popular girls. I didn't even notice her there in the chorus the first week. Then I did. It takes all my willpower just to keep my eyes off her. I'm an awkward face-in-the-crowd junior and don't have a prayer with someone like her. But I can dream.

I dream of overcoming my other big fear after public speaking: girls. They're a much bigger challenge for me, but I'll need to put off working on it for a while longer. I can only suppress one fear at a time.

Something amazing happens in the second week of rehearsals. Most of the other kids are busy with the director on stage. I sit and watch in the dark auditorium from ten rows back. Em approaches me quietly from the side, and I don't see her until she's right up next to me. She leans in and whispers the first words ever spoken between us: "Is this seat taken?"

Trying to look normal, I summon all my acting skills and whisper back, "Yes. It's saved for *you!*" She smiles and sits. For the next few minutes, I forget that I'm afraid of her. We begin a conversation and a relationship.

During the long weeks of rehearsals, Em and I meet whenever possible at the same pair of seats. These would be perfect opportunities to reach for her hand or give a little peck on the cheek, but I never even come close to trying. I think about it constantly, but I'm too timid to touch her. I've never touched any girl that way.

It finally happens at the cast party late on the show's closing night. Everyone's sitting around a big campfire behind the director's house. While all the others sing show tunes at the fire, Em and I hold hands back in the dimly lit periphery. We kiss. My heart practically beats out of my chest. I see colors on the insides of my eyelids. I'm deep inside the moment but also aware I'll never forget it.

Our romance progresses beyond the smitten stage. We learn about young love and stay coupled through the rest of high school.

Em graduates a year early. We work summer jobs, watch movies, hang out with friends, and prepare to move away to college at the end of August.

We enter the state university as freshmen. Sadly, within a couple of months, we join the ranks of high school couples unable to survive the transition to college. I initiate the breakup, but it hits me much harder than expected. I second-guess my decision every day, and I miss her. When spring semester comes, she transfers to a small state college a hundred miles away.

Weeks pass. We send letters sometimes, and her way with words makes me miss her more. I offer to drive up and meet her for a weekend in March. She agrees, and I make the trip on a Friday afternoon after my last class.

What used to be our romance has evolved into a best-friendship. We do lots of walking and talking during my visit. At night, I camp on the couch in the living room of the old house she rents with three other students. They're so welcoming and kind.

We talk about our classes and our schools' theater club productions. Em tells me of a casual relationship she's in, but I wave off hearing the details. I mention the closest thing I have to a relationship: anonymous girls in stairwells. That grabs Em's attention, and I clarify. An old music buddy from high school visits sometimes, and we try to meet girls by playing our guitars and harmonizing in the dorm's echoey central stairwell. It sounds great, and many prospects attend our impromptu concerts. But when the music ends, they all leave.

Em's going a different way with her extracurricular activities.

"Mostly, I hang out with this group from my church," she tells me. "They're such awesome people."

The church part's a surprise. She grew up in a Catholic family, but in our two years together, she never talked about having any religious feelings or expressed any guilt about our mutual devotion to sinning. So, I ask about her church and what her involvement with this group means to her.

"South Shore Christian Fellowship," she replies. Then, after a beat, "I'm born-again."

I've heard about the born-again Christians, but I'm ignorant about what it entails. So I ask, "What's *that* about?"

"It means I've had some really deep experiences. Now I see things more clearly. I live my life for Christ."

I'm conflicted. Back on my home campus, we call people who say these things "Jesus freaks." They're super annoying, always wearing their religion on their sleeves—and on their T-shirts, backpacks, dorm room doors, and weekly event flyers. I'm prejudiced against them for sure. But I respect Em, and I care for her so much that I rein in my disrespectful urges and ask more questions:

"What kind of experiences have you had?"

She hesitates a little and answers, "I've seen . . . like . . . miracles."

My eyes widen. I press for more. "What kind of miracles?"

"Oh, you know. Things that you think can't normally happen."

"Any specific ones that stand out?"

"Well, yeah. There's one that really affected me when I first got involved with the group. I don't know any way to explain it other than 'supernatural.' It made all my nagging doubts about the power of God disappear.

"It was a couple months ago, but I remember it better than yesterday. I'm with friends from church in the living room of Jory's house. He's one of the group's leaders. We order a pizza, and we're killing time 'til it arrives. At some point, I leave the room for a quick bathroom break.

"When I come back, one of the girls makes eye contact and pats the couch cushion next to her, so I sit down there. She asks if I realize one of my legs is shorter than the other. I answer that I don't, but thinking about it, I've had some pain in my back for a long time. Maybe it's from walking at a tilt?

"Jory chimes in and asks, 'Will you let us help you?' I think, 'What do I have to lose?' and say okay.

"He moves a couple chairs to the middle of the room, and we sit face-to-face. Then he leans toward me, looks at me with those eyes of his, reaches down behind my ankles, and lifts my legs until they're almost straight out in front of me. Everyone else is in a circle around us. It's like Jory's a doctor making his rounds surrounded by interns. And I'm the freaky case who makes their day.

"He looks around at the others and motions toward my feet with his eyes and his head. The others' heads all nod up and down.

"With my legs sticking out straight, my left looks at least an inch longer than my right. I mean, *noticeably* longer.

"Jory whispers, 'Prayer circle,' and the others all know exactly what to do. Then it starts getting kinda weird."

"Like it's not already?" I interrupt.

"I know, I know," she smiles. "Just wait."

"They start praying. Loudly. They're practically shouting at Jesus to heal me. Some of them pace around, mutter prayers, yell, then mutter some more. Others rock in place while they pray, their palms aimed up in the air. Everyone's in their own world. It's chaos, and it scares me a little. But it's also exciting with all the attention focused on me and my leg.

"After a few minutes, Jory takes over again. He yells over everyone else, really dramatic, like a televangelist." I watch as Em imitates him, pantomiming his trembling hands holding up her legs.

"He starts taking over the praying. Something like, *'In His Holy Name, with the blood of Christ, we beseech you O God. Heal her, Jesus! Heal her! JESUS! HEAL HER!'*

"Then Jory waves his hand, and everyone goes quiet. He gets all soft and solemn and says, 'Look.'

"Everyone looks at my legs. I don't see anything different, but the others' eyes are like saucers. They're already clasping their hands and thanking Jesus.

"Then I start to see it. My right leg slowly gets longer 'til it matches up perfectly with my left. It happened. I saw it with my own eyes. It was incredible."

"Did it hurt? Did you feel anything?" I ask.

"No, I don't feel anything except for the Holy Spirit and a lot of excitement. I'm smiling, maybe crying a little. Everyone's hugging everyone. Incredible energy.

"Anyway, as the excitement starts to wear off a little, I'm still the focus of attention and start feeling self-conscious. That's when *another* miracle happens, Barry!"

"No way. What?" I ask.

Em's eyes twinkle and she says, "Since we were on a roll, I prayed for that pizza to be delivered and, POOF! Hallelujah, just like that, the delivery guy knocks at the door."

BACKSTAGE

Em's membership in the church group didn't last the year. By the time she left, she wished she'd quit sooner. Her main reasons for staying as long as she did were the same as those which drew her to the group in the first place. She made social connections that made her happy and filled a spiritual cavity that had been aching since high school. The thought of giving up these benefits was daunting. When she did muster enough courage to voice her concerns about some of the group's practices, others quickly reminded her what she'd lose by leaving. Also, she was led to believe that she'd be upsetting certain key figures in the church who felt they'd made a spiritual "investment" in her.

She later found out that was just one of many lies she'd been told.

Emily had a *religious conversion experience*.[1] We've all known or heard of people switching religious affiliations. This is that, on steroids. It's an earth-shattering, life-changing, skydiving plunge without a parachute.[2] A spiritual house-cleaning. An emotion-charged reset for one's self-identity and sense of connection to a higher power.

Any big personal transformation has social consequences. It can strain or break family and friendship ties and leave others wondering what the hell happened. People in Em's old network were probably asking: *Was she privately questioning Catholicism and seeking another path to God? Was she gullible? Did she have a crush on Jory and let him have his religious way with her?*

But these possibilities all lay the responsibility on Em. They presume *she* decided she needed a change. Or *she* was too gullible to know what she was getting into. Or *she* gave up her power to get close to a charismatic man.

People who study coercion, exploitation, and abuse have a different view. They report that, if offered the right incentives, practically anyone can be recruited into an exploitative relationship. It's less about the person and more about the circumstances.

While indispensable for understanding human behavior, psychology's focus limits its field of view. The social sciences expand that view by focusing more on *relationships* and *contexts*. Rather than only keying on Em's personality, for example, we'd look at the shape and strength of her social network. At the time of her born-again experience, we would consider whether communication and moral support from her family and friends was faltering. It's normal to feel untethered after a major change in location and activities and to look outward for a soft place to land.

Another contextual possibility: Em's new church may have been proactively recruiting new members. Often this is done by encouraging or even requiring current members to befriend outsiders and convince them to attend social events and services. For Em, a college student detached from the friends and religion of her past, the attraction to the group must have been powerful. It offered exactly what was missing in her life.

Once a new member is "in," the group's incentives shift to retention and compliance. It tries to convince members that it supplies unique benefits. If successful, the members will believe there's no place they'd rather be. But if a member expresses doubts or acts inappropriately, the group may also impose punishments. The exact nature of the rewards and punishments varies by group, but they always advance pro-group behavior and discourage thoughts and actions considered detrimental to group integrity and harmony.

Love-bombing is an example of a positive incentive unique to social groups. A prospective or struggling member gets inundated with praise and affection. Whether spontaneous and sincere, or premeditated and strategic, the targeted individual is made to feel very special and is reminded they're unlikely to be so extremely appreciated anywhere else. The practice is especially common in close-knit groups that demand high levels of commitment. If you're looking to enjoy the short-term benefits of a good old-fashioned love-bombing, and you're not worried about the long-term social and psychological costs, Wikipedia.org lists some ripe possibilities. Look under "Cults."[3]

Much like people, groups usually try to make a good social impression by emphasizing their positives and masking their negatives. This is especially true when recruiting. But by the time a new member is around long enough to gain a fuller picture of the true rewards and costs, leaving may be very difficult. The most extreme example is the coercive group that prohibits contact with "unenlightened" outsiders, including relatives. Such groups unabashedly threaten potential defectors who would publicly air the group's dirty laundry. And they have an array of punishments—also called *sanctions*—used to keep current and former members in line.

On the milder side—say, if a member simply asks questions about a directive from the leader—the group may withhold rewards or verbally reprimand the offender. More serious offenses might trigger shunning, public shaming, fines, sleep and food deprivation, or detestable work assignments. When members dare to flee an oppressive group and go public, they may find themselves intimidated into silence with threats, harassment, smear campaigns, or frivolous lawsuits.[4]

Another perspective on groups imagines them as living organisms, trying to get by in sometimes harsh environments. Survival requires a steady supply of energy to carry out group functions. This energy could be in the form of people carrying out tasks. Or it may be converted from monetary resources into salaries, advertising, and so on. For the groups we're interested in, *human* resources provide much of the energy. A stable and supportive membership is the group's lifeblood.[5] If members lose interest or revolt and the group can't sustain its numbers, it can die. Losing members represents an *internal threat* to survival, like a serious illness or injury is to a body.

There are also *external threats*, such as competition from other groups. In Em's situation, there were probably multiple churches with youth groups, all vying for new members from the same limited population of new students. In another form of competition, my university financially supports a limited number of student groups. For a group to acquire this resource year after year, it must apply each fall semester and convince a panel of deciders that it's more worthy of support than other applicants.

Groups may behave badly when their survival is threatened. The group that finds ways to deal with internal and external threats has an advantage over those that don't. Unfortunately, this could mean adopting some unsavory practices such as self-isolation, deception, or violence. Few individual members would opt for such tactics. But when successful, they may become routine and rationalized through an ends-justify-the-means ideology: *Maybe our spiel to new prospects stretches the truth a little. But they'll be grateful to us later on for saving their souls.*

Many aspects of human behavior have causes or effects that only exist at the group level. An individual can't have a culture, hold an election, or be a social class without stretching the meanings of these phenomena beyond all recognition. Another example is the "wave" that moves around the crowd in a sports arena. The individual behavior involves standing with raised arms while perhaps shouting "Wooooooo!" Contributing to a wave makes people feel part of the group and adds to the event's excitement.

Juries can also be thought to have a life of their own, distinct from those of their members. For a particular case, every jurist may be absolutely *decisive* in their judgment of guilt or innocence. At the same time, the jury as a whole may be rendered *indecisive* or "hung" if one or more of those decisive individuals disagrees with the rest. As another example, a legislative body may broker into law a compromise bill that no individual legislator actually favors. If members on both sides—and most of their constituents back home—think it's "bad law," then why did it pass? The answer in four words: Groups are not people. To use a mechanical metaphor, a car behaves very differently from any of its constituent parts.

Between the individual and group levels is a vast world of social interactions where people can profoundly affect one another. *Social impact theory* helps explain how our interactions affect each other's beliefs and behaviors.[6] In Em's case, the theory shows that the group ritual made her believe wholeheartedly in a miracle she now knows never happened.[7] Here's how.

Social impact theory says that a person is influenced by others to the degree that the others are *(i)* numerous, *(ii)* higher status, and *(iii)* physically close. In Em's healing ritual, not one or two but instead around a half dozen friends presented a unified front, imploring her full participation. Em was the newbie, so the others all had higher status in the group. And they were physically close to her during the ritual. The theory predicts strong group influence under these conditions, compelling Em to see and feel what it wanted her to see and feel. And that's what happened.

Recently, I asked Em about her state of mind before and after her born-again experience back in college. She said, "It was like this . . ." and burst into a fake-ecstatic chorus of "Amazing Grace": "I once was lost . . . but now I'm found. . . .

"No, seriously," she continued, "back home, my mom's and dad's drinking was wrecking everything. At school, first-year classes in the

Nursing Program were killing me. And I felt *so* far away—in more ways than one—from you and all my other real friends."

The church promised Em one-stop shopping for salvation and community. Especially in the youth group, she found the comfort and belonging she craved. But, she went on to explain, she soon found herself paying a high moral cost, leading her to a crisis of conscience. To be part of the group, she had to pretend to be awestruck every time they suckered a newbie with the same leg-growing ritual they put her through. But by the time she saw through this and other deceptions, she was already so enmeshed in the group that it took months to summon the courage to cut ties and leave.

My teenage years included a growth spurt of six inches in a span of twelve months, mostly in my legs. Suddenly gangly, they regularly betrayed my sense of balance and dignity. But at least they sprouted at equal velocities. Em's unilateral leg-lengthening told a different story. While Nature's trickery took a month for each half-inch of growth in my legs, Jory doubled that result when, in mere seconds, he conjured for Em new bone, tendon, muscle, vascularization, nerves, and skin.

For as long as I can remember, something in me has resisted accepting miracles at face value. It's not that I ever made a conscious decision to be a "skeptic." Skepticism can happen simply when something seems too good, or too awful, to be true. *Scientific* skepticism, I've learned, happens when we hold up the claim to the bright light of scientific theory and evidence.

Nowadays, I think of miraculous claims as windows into the near-magical complexity of humanness. Em's story is a great illustration of how our brains, bodies, and social relationships can impel us to settle on a false or incomplete understanding of something that seemed inexplicable. But for me, it's the deeper, fuller explanations that inspire awe and fascination.

What do I mean by "deeper" explanation? Consider first its hypothetical opposite. A lone survivor of a fiery commercial jetliner crash is badly burned but lives.[8] When interviewed on the local news and asked why he thinks he alone survived, he says: "It was a miracle. I have no other explanation."[9]

The logical irony is that, by declaring his unlikely survival to be a miracle, he *is* offering an explanation. But it's a shallow one in the sense that the only support he offers for it is that he can't think of anything else. I wouldn't be so rude as to confront a gravely injured believer about his lack of logic or evidence, but for readers, I'll point out that the true explanation may simply be one he hadn't thought of.

"*We* can't explain it" might first appear to set the miracle bar much higher. Now it's no longer just one's own judgment at issue. Em was surrounded by people who fully backed and validated her belief. What more proof could you possibly need?

A lot, actually. If one person in the situation can't think of a better explanation, it's not a shocker if others with similar knowledge and backgrounds are also befuddled.

"Nobody can explain it" would be the strongest criterion—but the most difficult to implement. How can the average person know whether or not there's someone out there with a natural explanation for the seemingly supernatural phenomenon?

It's actually not as hard as you might think. Rather than reaching into your personal grab-bag of memories and experiences, or doing a Google search on "miracles," take a look at resources like the following, which are more likely to connect you to higher-quality information:

Snopes.com

SkepticalInquirer.org

ScienceBasedMedicine.org

QuackWatch.org

Skeptic.com

Any one of these may get you started on the path to scientific skepticism. You'll find virtually all the evidence that's been offered on behalf of hundreds of extraordinary claims. But additionally, the authors apply a skeptical lens and consider alternative evidence and explanations that are very much of this world—like a magician revealing how the tricks are done. The skeptical analysis is almost always more detailed, more systematic, more reliable, and *less* widely disseminated than the original claim. Because news media gatekeepers and the public prefer to circulate extraordinary claims, ordinary explanations are usually consigned to the shadows.

How did Jory make Em's leg grow?

By now, you've probably realized it was a trick. We like magic tricks because they make the mundane look miraculous.[10] Conjurors purposely deceive you and tell you they're doing so. That makes them "honest liars,"[11] working within the laws of physics to make it appear like they're not.

But when a trick is performed in a religious setting, the audience isn't there to be fooled for fun. The subject is primed to see divine intervention. The job of the facilitator (clergy, shaman, youth group leader, etc.) is to make them believe their own eyes. Since willful deception is involved, it's fraud, pure and simple. It makes you wonder if the Jorys of the world really believe this is a legitimate answer to the famous question, WWJD? What would Jesus do?

The leg-growing stunt in particular has a history as old as faith healing.[12] The first step is the dramatic build-up. In Em's case, Jory literally set the stage, then cued his cast to pray for Em with escalating frenzy. The attention and chaos made it exciting for her. And being emotionally jacked-up and distracted helped her suspend any skepticism she might have felt. Plus, she was under a lot of pressure not to disappoint the group.

As for the trick itself, there are at least three ways it's usually done, and they can be implemented individually or in combination.

1. *The Shoe-move.* Most shoes leave a little wiggle room, and Jory could capitalize on this fact. When he first grabs Em's heels, he *(i)* pushes the heel of her right shoe tight against her foot to "shorten" that leg

and *(ii)* gently tugs the heel of her left shoe to "lengthen" that leg. This creates the illusion that her right leg is shorter than the left.

At the moment of truth (oh, the irony of calling it "truth"), Jory misdirects everyone's attention to Em's supposedly growing right leg. In fact, he's slowly pushing the heel of the left shoe back where it belongs, eventually aligning it with the right. Hallelujahs and high fives, the legs now match! Of course, Jory *could* have claimed he shortened Em's left leg to match her right. But, as an opportunistic punster might say, that would seem a lesser "feat."

2. *The Ankle-turn.* This method is simple, but it works. Grasping Em's feet, one in each hand, Jory starts by ensuring the toe of Em's right shoe points a bit down toward him, while her left toe points up to the ceiling. This makes her left heel appear to jut out more than the right. He can mask the different ankle rotations by splaying her feet and moving them apart, making them harder to compare. Pretending to verify the disparate leg lengths, he briefly brings the legs together and taps the heels together. Jory could then gradually realign the shoes by pivoting the ankles. After he finally confirms the miracle with a couple more heel taps, Jesus returns to the business of answering prayers about sporting events.

3. *The Hip-scootch.* Standing up straight with knees locked, most people can easily lift either foot an inch or so off the floor. All it takes is a slight shift of the hips. The same move can be done seated with legs extended. A little scootch of the hips, and one leg draws in while the other pushes out. Try it!

   When Jory first puts his hands on Em's feet and gets comfortable in his chair, he moves her legs a few inches to his left. That's all it takes to shift Em's hips and make her left foot appear shorter than the right. As the ritual builds to its climax, Jory slowly shifts Em's legs to his right, realigning her hips, and making her left leg seem to grow to match her right. *Voila!* The group has managed to make their all-knowing and all-powerful creator (a) realize he messed up Em's leg and (b) fix it.

The fact that these are simple tricks doesn't mean they're easy to spot. They're effective because they deceive so reliably. When Em told me about her conversion, I would have been no better prepared than she to see through the ruse. Still, I'm sure that it would have affected us differently. Though I felt spiritually content at that time in my life, Em wanted a miracle and got one.[13]

During the time Em counted herself among the ranks of the born-agains, I kept my distance—or maybe she was keeping hers. But eventually we reconnected and talked about her conversion experience. In the interim, I'd read about how the trick was done, along with a number of other deceptive faith-healing practices.

"You realize that was a sleight-of-hand trick, don't you?" I finally got to tell her a few years later.

She said, "I know. I'm not with them anymore. They betrayed my trust, and I felt stupid for letting them."

"Yes, they did," I told her. "But no, you weren't stupid. You just didn't know at the time how the trick was done."

Decades later Em still hasn't completely let go of her resentment toward that group or the pain it caused her at a vulnerable time in her life. She's still religious but would never abide by the use of the deceptions used on her, even if they're a means to glorious ends.

Jory had a different moral code. He knew exactly what he was doing in Jesus' name. I wonder how many legs he's pulled.

CHAPTER NINE

Astronomy

I don't believe in astrology; I'm a Sagittarius and we're skeptical.
—Arthur C. Clarke[1]

FRONTSTAGE
Astrology works!

This is according to two confident, intelligent young students eager to enlighten me on the subject.

One is a fellow senior at UMass/Amherst. We meet through a project I'm doing as part of my honors thesis in psychology.[2] It's a lab experiment with human subjects. There's serious science happening, but my secondary objective is to play with all the equipment packed into the small control room: a polygraph machine, video recording decks, cameras and microphones, audio systems, and other toys of the trade. Nerd heaven.

Enter Wanda.

After setting things up, practice trials proved that another pair of hands would be helpful at certain moments in the study. I ask my advisor if we could get a research assistant and compensate them with course credits. He agrees. I put an ad in the *Daily Collegian*, and Wanda emerges as the most qualified and motivated applicant. A week later, we start collecting data.

Each session of the experiment is ninety *long* minutes, of which Wanda and I are only busy for fifteen. Instead of zoning out to the hum of electronic equipment and clicking polygraph pens, we chitchat. It's mostly superficial. Hometowns, families, relationships, classes, dorm life, campus events, the whole "UMass/ZooMass" experience. She plans to get a master's in speech pathology. I'm just starting to think about grad school. We talk about science, which we both love and in which she's significantly better trained at this point.

Wanda brings up astrology around the third day. She believes in it sincerely and seems to know a lot about it. In fact, astrology is as much a science to her as astronomy or biology. It sounds kind of beautiful—mathematically charting the patterns formed by planets against the backdrop of constellations, applying time-tested inferences from these patterns to the human experience. Wanda's mostly into sun sign astrology and especially likes Linda Goodman's books. Horoscopes speak to Wanda as clearly and accurately as she speaks to me.

"Seriously, Barry, it works."

"Why do you think it works?" I ask in a tone more curious than skeptical.

"The horoscopes always fit so perfectly—me, my friends, my mom and dad. Okay, they're not always perfect, but they always seem to fit at least *really* well. Like when I met you, within five minutes I could totally tell you're a Libra, right?"

"I'm a Leo."

"Ohhhh, okay. Yeah. Leo was probably my next guess."

"Actually, I'm an Aries."

"Stop that!" she laughs.

"Sorry. I'm really an Aries. What's your sign?" But I immediately stop her before she can answer. "Wait. I'll bet *anything* you're a Capricorn!" She looks stunned, as I'd hoped. She indeed forgot she'd mentioned her late December birthday when I interviewed her two weeks ago. I let her be amazed a few more seconds before admitting my cheat.

"Seriously, Wanda. How can the stars and planets at the time you're born decide all these things for your whole life—personality, jobs, relationships? How could that even work, scientifically?"

"I don't know," she admits. "But it's been used thousands of years so there must be something to it. I'm sure there's an explanation. For me it's like that TV." She points to the monitor where we can view the subject in the next room. "I don't know how the thing works. It just does."

As the project moves along, astrology is a regular topic of conversation. We have friendly arguments interpreting our horoscopes printed in the daily paper. I'm still skeptical despite Wanda's enthusiasm and the cultural clout of Linda Goodman's thirty million books in print.

It's six years later. I'm a new professor at the University of Iowa and teaching an honors social psych course. A dozen ultra-smart students and their anxious prof are seated around a conference table in a seminar room. I have imposter syndrome, and I have it *bad*. I'm younger than some of the students, barely older than any of them, yet accepting a salary to teach them. I respond by overpreparing, but it's never enough.

We start with units on scientific theory and research, soon confronting the following question: *What makes science different from other ways of knowing the world, such as through poetry, personal experience, or common sense?* We discuss some of the ideas offered by the students, then talk about contrasting a real science with a field that pretends to have that status: a *pseudoscience*. I illustrate this using astrology, something I've learned more about since my days back at UMass.

Few, if any, working scientists consider astrology to be a science. Most astrologers, and a third of the public, believe it is. I don't dwell on it in class, but I do point out that astrology lacks a coherent theory, has no supporting evidence, and contradicts established fields like physics, astronomy, and physiology.

Celeste, one of the most brilliant students I'll ever know, sits there taking it all in. She approaches me after class.

"Hi Celeste!" I greet her as she sits near me. We've not spoken, but her name jumped out at me when I checked attendance at the start of class. It sounds like *Celestial*—"of the stars and planets."

"Hi professor. You said that astrology isn't a science . . . but it is!" The way she said this was more zealous than defensive. "I learned a lot about it from my mom. She's a professional."

"A 'professional'?"

"Astrologer."

"Really?"

"Why would you say that astrology isn't science?"

"Like I said, lack of theory and research." I turn the question around. "Why do you think it *is* a science?"

Without hesitating, she answers, "There *is* research to support it, and it's actually the oldest science. The mainstream's just too closed-off to accept it. Besides, it works! I've seen proof."

"I don't get how it can explain its cause-and-effect." I have to ask, "How do objects so far away define our whole lives based only on the instant we're born?"

"I don't think anyone knows for sure." She *up-talks* her next sentences like the teenager she still is. "Maybe gravity has something to do with it? The sun and the moon affect ocean tides, right? Humans are, like, 65 percent water? So maybe that's the mechanism?"

"Okay. You said you've seen the proof. What is it?"

"Well . . . there's tons! Like, last month my mom gave me a chart she drew up for my boyfriend." I'm so glad she dropped the up-talk. "I called him up and read to him over the phone what mom had written based on the chart. They'd never met before, but it was *really* accurate. We were both totally blown away!"

"Interesting." Then I had an idea. "Do you think your mom would let us test her?"

She looked surprised, thought, smiled, and said "Maybe."

Backstage

The Urban Dictionary jokingly defines *astronomy* as the study of astronomers. I have an alternative: *The deformed spawn of astronomy and astrology.*

A quarter of Americans are unwitting astronomers, confusing these two different pursuits.[3] The reasons for the popular mash-up are obvious: They sound similar, they both look at things in the sky, and both are touted by experts who seem to know a lot of stuff. Unless you proactively learn the differences, these superficial similarities tend to stick in the mind.

There's an old joke that the real difference between astrology and astronomy is fifty IQ points. Not so. There's hardly any connection between intelligence and belief in astrology.[4] This is never more evident than in the many who understand the basic distinctions but still like astrology. It's not about intelligence. It's about uncritical thinking.

Wanda and Celeste are cases in point. Both are educated and science-oriented. Both know science pooh-poohs astrology. But neither knows about informed scientific critiques. There's no debunker out there approaching the huge impact of best-selling author Linda Goodman. Plus, scientists are bluntly dismissive, making it easy to dismiss them in return. Individual scientists indeed are flippant about astrology. But science as a whole has taken it seriously. There are tests.

Astrology either can do what it claims, or it can't. If it really can predict personalities, relationship compatibility, and careers, then it definitely qualifies as an extraordinary claim since we know of no way this can happen. In fact, it would send a number of active, long-standing scientific fields back to their whiteboards to work out how to integrate astrological forces into their models. But sciences are loathe to retool unless they really, truly must. While they're open to new possibilities, evidence must be rock solid before casting aside knowledge developed through many years of rigorous testing.[5]

There is some research claiming to support astrology's claims. One study had over twenty-three hundred subjects and was published in a major social psychology journal. Its authors predicted a correlation between an introvert/extrovert personality test and zodiac signs. This should be an easy win for astrology: Six zodiac signs predict introversion, six predict extroversion, and psychologists have very reliable introvert–extrovert personality tests.[6] The results did, in fact, support astrological predictions: There was a significant correspondence between sun sign and score on the introvert–extrovert test.[7]

Vindication for astrology? Not so fast.

The authors and the journal's peer reviewers missed a serious flaw. We all tend to see ourselves based on feedback received from others—*reflected self-appraisals*. Believing these appraisals, we act on them in a so-called *self-fulfilling prophecy*.[8] If true in this study, subjects may have responded according to their signs if they already believed their signs made them introverted or extroverted.

Later studies proved that was the case. If respondents don't know what traits are associated with their zodiac signs, their introvert–extrovert scores don't match their signs. More revealing is when subjects are given false information about traits associated with their signs. Tell an Aries they're introverts, for example (their horoscopes actually say the opposite), and their introvert–extrovert tests will correspond with the false information.[9] Results such as these deal a major blow to sun-sign astrology.

In a more famous study, Michel Gauquelin published evidence for the *Mars effect* in 1955.[10] He tested whether being born with Mars in certain parts of the sky made people more likely to be sports champions. He found 22 percent aligned with the Mars prediction—more than 4 percent above chance. It's not a huge effect, but it would be good news for astrology if true.

It's not.

Several major follow-up studies were conducted. None reproduced the Mars effect. The more careful the methodology, the smaller the effect. The most recent attempts to replicate Gauquelin's findings show no effect at all.

One of the problems was the lack of a clear definition for "champion." There's evidence that Gauquelin capitalized on ambiguity when, already knowing their birthdays, he included or excluded athletes in his study based on whether he alone considered them to be *true* champions. It biased the results in his favor.[11] We don't know if he purposely cheated, but the follow-up work is a testament to the scientific *norm of replicability*. Especially with extraordinary claims, it's critical that results can be reproduced by other scientists, even those who might be skeptical.

Many other studies tested astrology's claims about personality traits, sign compatibility, occupation, medical issues, or physical characteristics. None of the predicted effects were found. One study published in *Nature* stands out from the rest, not only for the prestige of the journal, but also because it was conducted by a skeptic in cooperation with astrologers.[12] Physicist Shawn Carlson collaborated with the National Council for Geocosmic Research and only used well-reputed astrologers in his tests. He focused on a central astrological claim:

> The positions of the sun, moon, and planets at the moment of birth can be used to determine the subject's general personality traits and tendencies in temperament and behavior. [p. 419]

In the first of two studies, Carlson's subjects submitted their birth data to him. He distributed it to the astrologers, who produced natal chart interpretations—which I'll just call "charts"—for every subject. For the test, subjects had to select their own chart from a set of three: theirs plus the charts for two others chosen at random. If only guessing blindly, about one-third of subjects should choose their own chart. Astrologers claimed that *at least* half of the subjects should be able to select their own chart—a very modest criterion but enough to be statistically valid. In the end, almost exactly one-third chose correctly, no better than chance. Subjects couldn't recognize their own charts.

Astrologers were the subjects of the second study. They'd claimed to be able to recognize correspondences between a natal chart and results from a personality test called the California Personality Inventory. Astrologers were each given three subjects' natal charts and the personality test for one of those three subjects. The task was to choose the natal chart that went with the personality test. Astrologers predicted they'd get at least 50 percent correct compared to the chance result of one-third. The actual success rate was again almost exactly one-third.

To summarize:

- Subjects can't pick out their own astrological charts.
- Astrologers can't match subjects' personality test results with charts.
- The more stringent the test, the less astrology's stars align.

With Celeste as a go-between, I gave her mom our instructions:

1. Perform a natal chart reading for one subject based only on his or her birth information.

2. Using that reading, complete the attached Personal Traits Checklist for the subject. The test includes pairs of opposite descriptors: organized, disorganized; composed, excitable; conservative, liberal; etc.

3. The subject will also complete the checklist.

4. Random guessing will result in about a 50 percent match between the two checklists. Matching at 80 percent or greater will be considered a successful astrological reading. Less than 80 percent will be deemed unsuccessful.

5. We will report your results after both tests are submitted and compared.

She accepted the conditions, no questions asked.[13] Celeste conveyed that her mom was sure she'd score close to 100 percent. But the following week, the mom asked her daughter to pass along a question to me. Celeste said, "She says it would help her to know if the subject is male or female. Can you tell her which?"

She's right, it would help. Demographic categories like gender correlate with other traits. But she'd already agreed to use only birth information, so I had to tell her no. Celeste already knew the answer and replied, "I didn't think so."

A few days later, Celeste received her mom's completed checklist and brought it to my office.

"Is everything okay?" I asked. She looked like she wanted to tell me something.

"Yeah, well, when my mom gave me the checklist, she said it may not match the subject's very well. She said it's because astrology knows people better than they know themselves."

Celeste immediately knew that her mom had broken the agreement by rejecting anything other than a successful result. Celeste had remembered from class the norm of *falsifiability*: If there's no conceivable way to disprove a field's claims, then it's not a science. Mom had made her astrological reading unfalsifiable. A positive result would support astrology, but a negative result could not refute it. To Celeste, it revealed astrology's true pseudoscientific colors.

With the test effectively nullified, I asked, " Have you thought any more about the time you read your boyfriend's horoscope to him over the phone? How did it fit him so perfectly?"

"I figured out what happened!" She perked up. "For one thing, sometimes I talked to my mom about him, so she kinda knew him already. But also, I didn't actually read him everything on the chart during the call. I just read the parts that fit him best." Celeste had unconsciously produced a *confirmation bias*. Like her mom, she'd unknowingly downplayed any disconfirming information.

The test results actually matched at a rate of 47.5 percent, about what chance would predict. Had I been more sensitive, I'd have anticipated the potential of our little test to create some mother-daughter friction. When eventually I realized this, it seemed best not to risk making things worse by giving an A+ student's mother an F. I never told them the results. They never even asked.

Coincidentally (I think), I received a surprise email from Celeste a few months ago. I'd not been in touch with her for decades following that semester's end. She talked about her family and successful law career and memories from college including my class. In my reply, I mentioned our project, and she also remembered it well. She told me she still dabbles in astrology, but not as a True Believer. When I asked about her mom, she said they have some different views, but it doesn't affect their special bond. Nor, said Celeste, did her mom's concerns about her daughter's astrologically incompatible fiancée deter them from marrying. It's been thirty-three starry-eyed years and counting.

Astrology is what happens when a connect-the-dots game is taken too seriously. Our Western version came from constellations that ring around the horizon. Their meanings were assigned twenty-five hundred years ago through the cultural lens of Babylonian stargazers. Connecting stars to form the Taurus constellation outlined what looked like a bull to them, so people born with the sun in Taurus were declared bull-like and prone to anger. They thought Libra looked like scales, so Librans were balanced, and so on. The sun, moon, and five visible planets had their own traits, too. They move through the constellations at certain times of the year, with their "interactions" providing endless fodder for storytelling.

Constellations are not like connected dots on an overhead dome. Their stars are distributed in three-dimensional space, at greatly varying distances from us. Viewed from other vantage points in our galaxy, the patterns we see from Earth don't even exist.

Neither are the constellations self-evident, even from Earth's vantage point. They arise from cultural symbols with significance to people living in particular regions at particular times in history. Other cultures see other constellations. What is shared across cultures is the human need to quell randomness by inferring patterns and meanings.

Most followers of astrology aren't aware that, during the two thousand years of the existence of the zodiac, constellations have shifted a tenth of the way around the horizon. Aries are still considered as Aries, for example, despite the sun now being in Pisces at the time of their birth. Some astrologers have tried to adapt to this, but no coherent theory impels them to fix the problem, or directs them toward solutions.

The lack of a rigorous, testable theory is a problem that can't be overstated for a field aspiring to be scientific. *How* do planets and stars impact our bodies and our brains? *Why* does being born a minute earlier or later alter one for life? If it were a science, much activity would be aimed at opening astrology's black box.

Speaking of the moment of birth, why don't astrologers ever question this assumption? If astrological forces are powerful enough to span light-years reaching the emerging fetus, undeterred by Earth's atmosphere, trees, buildings, or hovering obstetricians, then how could they be thwarted until the moment of birth by a mere abdominal wall? For that matter, "the moment of birth" is also problematic. It's common for minutes to pass, or even hours, in some cases, from the first crowning of the baby's head out of the birth canal to the completed delivery. Every minute counts in an accurate natal chart, so when exactly *is* that moment of birth? Having an actual theory would help leverage an answer.

Celeste's metaphor about tidal effects is often raised as a possible mechanism. But—forgive me, Aquarians—it doesn't hold water. Tidal forces are extremely weak. Across an ocean thousands of miles across and thousands of feet deep, the moon's changing tidal force causes the sea level to change about twenty-four inches. Scale the ocean down to a human-sized sack of water and the moon's tidal force reduces proportionately. It becomes much less than that of a mosquito landing on the back of a neck. If astrological forces are tidal, then the horoscope would have to account for the impacts of far greater tidal forces: alighting insects, people and objects in the birthing room, the birthing room itself, vehicles passing outside, and anything else within the slightest gravitational reach of the just-emerged fetus.

There'd still be the question of how in heaven's name the minuscule tidal sloshing of fluids in the newborn makes her forever an extrovert or predisposed to be an accountant.

Astrology has existed more or less in its current form for twenty-five hundred years. Its longevity might appear to validate its status as a science. The opposite is true. No legitimate science remains static for long. All have cutting edges with active research programs. All evolve as their theories sharpen and their data collection improves. Staying unchanged for centuries is a big red flag commonly seen flying over pseudosciences.[14]

Astrology's success isn't as a science but, instead, as a popular, self-perpetuating belief system. People adopt it for many reasons.[15] Psychologically, it connects them to something that feels transcendent. It's also fun, self-validating, and a coping mechanism that can reduce stress and uncertainty. Socially, it's sustained through websites, organizations, networks, mass media, and readily available practitioners. Today, there's far less stigma attached to believing in astrology than there is for nonstandard religions and many other paranormal beliefs.

I would sometimes use a classroom demonstration to show how easy it is to be taken in by pseudoscientific claims.[16] I'd hand out printed sheets to the class with an official-looking "CompuScan Project" logo and instructions for the take-home exercise. The purpose, students were told, was to help me fine-tune a computerized personality test. They were asked to supply their name, birthdate, birthplace, favorite color, favorite number, and a brief description of a recent dream. Astrology wasn't mentioned.

A week after turning in their forms, I gave each student a personalized sheet displaying the information they'd given and the main "CompuScan Results." For example,

You have a need for other people to like and admire you, and yet you tend to be critical of yourself. While you have some personality weaknesses you are generally able to compensate for them. You have considerable unused capacity that you have not turned to your advantage. Disciplined and self-controlled on the outside, you tend to be worrisome and insecure on the inside. At times, you have serious doubts as to whether you have made the right decision or done the right thing. You prefer a certain amount of change and variety and become dissatisfied when hemmed in by restrictions and limitations. You also pride yourself as an independent thinker and do not accept others' statements without satisfactory proof. But you have found it unwise to be too frank in revealing yourself to others. At times you are extroverted, affable, and sociable, while at other times you are rather unrealistic.

In fact, half the class received this exact description. The other half received an oppositely worded version. So, where the above says

At times you are extroverted, affable, and sociable . . .

the reversed version says

At times you are introverted, quiet, and introspective . . .

Finally, there were two questions at the bottom of the page:

1. How well does the *CompuScan* analysis describe your specific qualities?
   PERFECT   EXCELLENT   GOOD   FAIR   POOR   NOT AT ALL
2. How well does the *CompuScan* analysis describe your general qualities?
   PERFECT   EXCELLENT   GOOD   FAIR   POOR   NOT AT ALL

Since the personality descriptions were totally generic, we might expect an occasional FAIR or GOOD response but rarely EXCELLENT or PERFECT.

That's not at all what happened. For both versions, responses were about evenly split between EXCELLENT and PERFECT. Answers of GOOD or below were rare, sometimes nonexistent in a class of thirty.

This is the *Forer Effect*—a kinder tag than the often-used *Barnum Effect*, as in P.T. Barnum's "There's a sucker born every minute." It's not about gullibility. It's more about our pattern-seeking minds trying to make sense of things. The more wishy-washy the description, the easier it is to make it fit. In fact, we're complicated. Different facets of our personalities activate in different situations. For example, the same person who dominates at work may be docile with family. When it comes to the Forer Effect, it's the subjects doing the work of making the description fit. They reach back for confirming memories, creating the biased impression that the vague description fits them to a tee.

Wanda and I stayed in touch after graduation and have remained lifelong friends. Preparing this chapter, I was all set to ambush her with a phone interview about her ongoing interests in astrology. We hadn't touched on the subject in many years, and I presumed that was to avoid conflict.

I had ten questions prepared. I called her up and asked the first: "Wanda, how would you describe your current beliefs in astrology?"

"I don't believe in it at all!" she insisted. Suddenly, my next nine questions were irrelevant. I had to improvise.

"Why not?" was all I could come up with under pressure.

"Because of you, ya bum!" she chuckled. "It was after you told me about all the research that's out there."

This sounds like a "just-so story," but the quoted words are genuine, as was my surprise at her skepticism.

After college, I learned something about people with Wanda's sun sign that I don't think she knew. It's maybe even a little hurtful: Capricorns are supposedly gullible.[17]

But she can be proud of her skeptical turn. She's a living falsification of sun sign prescriptions. She's her own woman, free of stellar predestination. The anti-Capricorn.

CHAPTER TEN

You Made the Earth Move

To minimize loss and damage in a quake, try not to own things.
—The Onion[1]

Frontstage
Five seconds into it.
"Whoa, you feeling this?" Matt asks in a shaky, higher-than-normal voice. I pause and concentrate.
Ten seconds.
"Definitely. Earthquake!" I answer.
"No friggin' way!" Matt's laughing nervously.
"Yup," I say. "Just like I told you, right?"
Fifteen seconds.
I'm wondering: *Should I ride it out under the desk? Do I make a run for it into the corridor and out the front entrance?* We each glance quickly around our small windowless office. He looks at me. I look under my desk. He looks at the door. I look at *him*. He looks under *his* desk. *I* look at the door.
We can see through the door's sidelight into the office hallway. Nobody's evacuating. Does that mean it's safe? Or are we all doomed because they're waiting to see if others leave? I don't know. I'm not from around here. I remember hearing somewhere that you're supposed to go into the bathroom because the pipes reinforce the walls. Wait. That was for tornadoes. I think it's doorways for earthquakes.
Thirty seconds.

I lean down from my chair, angling toward the space under my desk. At the same moment, Matt makes his move toward the door.

The shaking has stopped, but I don't know for how long because my pounding heart was also shaking me. We freeze, holding our breaths.

Thirty-five seconds. Forty. Forty-five.

It's definitely over. We exchange grunts about what just happened, then wander into the corridor where others are starting to mill about.

I'd only been in California a few weeks before the earthquake. It was early September 1988. I'm at Stanford on a year-long research fellowship. Matt and I are both in our early thirties and both on leave from our home universities. Our office assignment was random. We have no research interests in common. Nor is there any overlap in our personalities. We've yet to have an interesting conversation. He tells me he feels (*feels!*) the scientific method is optional in social science research. (Social *science* research!) I feel that his feelings on this issue are irrelevant.[2,3]

I mostly ignore Matt. I'm being paid to do my own research for an entire year in one of the best academic environments in the world. I have no teaching obligations and no committees. I've hit the ground running, going at it for long hours and getting stuff done. I love my job, and it doesn't feel like work.

Let's go back a few hours. I arrive at the office about 8:30 as usual. Matt shows up just as my coffee machine churgles at the end of its brew cycle. I offer him some. He accepts but first takes his mug out to the men's room to rinse out yesterday's remains.

When he returns, I spin my chair toward him. I think *Why not?* and decide to try talking with him. I take a sip from my mug and say, "You're not gonna believe this. Last night I dreamt we had an earthquake. It was *so* real. That ever happen to you?"

"Nope," he replies, turning to his desk and flipping open the morning newspaper. I'm left sitting there facing the back of his chair, wondering if I'd offended him somehow. I can't think of anything I did. So I let it lie and decide he probably grew up in an orphanage where he was beaten for talking, or for liking science, or both.

He never thanks me for the coffee.

The earthquake arrives four hours later.

That evening, I learn from the local newscast that it lasted ten seconds, with no significant damage and no injuries reported. Only ten seconds. Sure fooled me.

Since it happened, I was amused by the idea that my dream predicted the earthquake. I also have this vague feeling that somehow I *caused* it. I don't believe that, of course. It's strange the thought would even occur to me.

Backstage

How remarkable was it for my earthquake dream to coincide with an earthquake? Was this *just* a coincidence?

Often when the story of an extraordinary event is told, some parts get emphasized or *sharpened* and other parts are downplayed or *leveled*.[4] This can happen because the storyteller's memory has enhanced some story elements and discarded others. It can also happen as a series of unconscious decisions made on the fly. Telling a story in a social setting, it's natural for the teller to emphasize the parts that seem likely to be most interesting to listeners and to gloss over whatever might seem boring or irrelevant. It feels good, after all, to have people interested in what we have to say.

These effects are amplified when a story moves through a social network. It's like the "telephone game" where a message is whispered from person to person around a circle. By the time it returns to the starting point, it's been completely altered by incidental "noise" along the way. Sharpening and leveling also distort the message, and the noise compounds with each retelling.

This is a big problem if you sincerely care whether or not a claim is true. The sharpened parts of the description may include made-up "facts" that mislead listeners. The leveled parts are gone forever, though they may have held the key to knowing what really happened.

The main events in the earthquake story are true. One night, I dreamt there was an earthquake; later that day, there actually was one. But for the sake of making a point, I admit that I sharpened with reckless abandon.

I overstated how frantic we were while the quake was happening. It wasn't our first rodeo.

I don't have a clue when I actually arrived at work that day, but stating an exact time makes me sound detail-oriented.

I certainly didn't time the events down to the second. Who does *that*?

And most egregiously, I only told Matt about the dream after the quake, not before. Had I been assigned to an office with someone I actually enjoyed talking with, I'd have been a lot more likely to mention the dream when I first saw him.

As a side note, the timings I reported approximate my subjective experience of that brief quake. It really seemed much longer. Also, my officemate's unpleasantness (but not his real name) was portrayed accurately. And he *never* thanked me when I gave him coffee.

As for leveling, I left out several *very* important details relating to the question of whether the coincidence was truly remarkable.

The Stanford University campus is only a couple of miles from the San Andreas Fault. It's also in easy striking distance from the more dangerous Hayward Fault. There are thousands of measurable earthquakes every year in the area, most under magnitude 3.0 and felt by few if any people. But when the epicenter of a magnitude 3.0 earthquake is close by, it's definitely "feelable"—a term used by the locals.

During that year at Stanford, I'd estimate there was about one feelable quake per month. That's not enough to become habituated to them, and so my heart would flutter and adrenaline would rush every time. The particular quake in the story was only a 2.5 magnitude, but its epicenter was only two or three miles away on the San Andreas.

Another leveled fact is that I had *a lot* of earthquake dreams that year. I'd lived in the area before—five years spanning grad school and a post-doc fellowship. I knew what to expect, and I started dreaming about earthquakes months before my return.

Intuitively, you can guess that the more earthquakes there are in a year, and the more earthquake dreams you have in a year, the more likely you'll experience the kind of coincidence I described. But intuition doesn't help all that much in terms of estimating the actual likelihood of the coincidence. For that, I ran the numbers. The results may surprise you. I've relegated the calculations to the endnotes for the one or two readers who may be interested.[5]

First, a quake per month makes for a 3.3 percent chance there's a quake on any given day.

Second, let's say for argument's sake I had two earthquake dreams per month—probably a conservative estimate. That's a 6.6 percent chance of an earthquake dream on any given day.

Third, multiply those two numbers to get the likelihood of *both* the earthquake *and* the dream on the same day. It's only about one in 450.

Fourth, apply Murphy's Law: If something *can* go wrong, eventually it *will*. In our case, if there's *some* chance for an earthquake coinciding with a dream, eventually it will. But how likely is it across a year's worth of opportunities?

Pretty likely, according to the calculations. There's a 55 percent chance the coincidence will happen at least once in 365 days. Another way to look at this: If you had to bet, it would be wiser to bet *for* the coincidence than against it. Coincidences don't have to be unlikely. This one was more likely to happen than not.[6]

What about that funny feeling I had that, by dreaming about the earthquake, I *caused* it to happen? It was striking because, on the one hand, I knew in my bones that I didn't cause the earthquake. But on the other hand, I was aware of my brain wanting to "go there." What was that all about?

One possibility is simply that my memory associated the earthquake coincidence with a book that I read and enjoyed a decade before. Published in 1971, Ursula K. Le Guin's classic *The Lathe of Heaven*[7] tells the story of a man who discovers that his dreams alter reality. Not just his own reality. Everyone's. Needless to say, hijinks ensue. It was such a memorable premise for a sci-fi novel that I may have sensed a dim connection between the book and the earthquake coincidence, without it quite rising to a conscious level.

There's a deeper explanation that's more likely the correct one. It involves the concepts of *patternicity* and *agenticity*.[8] Patternicity is the brain's tendency to infer a pattern in a set of stimuli, whether or not the pattern is there. An example is seeing constellations or "signs" in the stars. Another is the "gambler's fallacy." For instance, a roulette player notices the wheel spun "red" the last three times, and so he bets "black" in the belief that it's overdue. It's not. Each spin's result is independent of what's happened before. But probability theory predicts that you'll sometimes get strings of identical spins by chance alone.

Agenticity is the tendency to infuse patterns with meaning, intention, and agency. Astrology associates human traits with each sign and claims that people born under a given sign will have its traits. The roulette player, upon betting "black" and losing, may decide the wheel "has it in for him."[9]

Against my better judgment, my brain inferred a pattern connecting two prominent events: "earthquake dream" and "earthquake." The basis for this was that they both involved earthquakes and roughly coincided in time. There's a reason our brains evolved to infer such connections: Noticing them has great survival advantages when it comes to, say, harvesting resources or avoiding dangers. Of course, sometimes these inferences lead us to infer dangers that aren't there. But better safe than sorry.

My brain didn't merely connect the dream with the event. It inferred agency: My brain wanted a reason for the earthquake; my dream had just preceded it; therefore, my brain decides, my dream must have caused the earthquake. Agenticity is baked into our neurons. The brain would rather leap to false conclusions than suspend judgment while awaiting better information.

We can't stop our brains from doing these things. We can't sidestep evolved, pre-wired tendencies just by knowing about them. But what we can do is reflect on them, recognize them for what they are, and not accord them undue significance. Your gut feeling isn't always reliable. It pays to apply this kind of thinking in any situation where you notice a coincidence and feel that it may have been extraordinary. Often it wasn't. What *is* extraordinary, as far as I'm concerned, is that we can use our brain both to reflect on these things and to compensate for its limitations.

I returned home in mid-August 1989 after a very productive and satisfying post-doctoral year at Stanford. Two months later, on October 17 at 5:04 p.m. Pacific Time, the San Francisco Bay area was struck by a magnitude 6.9 earthquake—twenty-five thousand times stronger than the one I embellished in my story. Its epicenter was Mount Loma Prieta on the San Andreas Fault. Sixty-three people died, nearly four thousand were injured, and there was over $10 billion in damages. The Stanford campus sustained damage to two hundred buildings.[10] This wasn't the dreaded "Big One" that geologists warned was coming. But it was bad enough, and I never dreamt anything so terrible could happen.

CHAPTER ELEVEN

Dowsing the Dowser

Now there are so many scientists who believe in dowsing that the suspicion comes to me it may only be a myth after all.
—CHARLES FORT[1]

FRONTSTAGE
A man unpacks his soft-sided suitcase and lovingly arranges eight or ten strange objects on long tables up front. We're in a cheery classroom-like space at the Iowa City Public Library. He's come all the way from rural Grand Junction in the western part of the state. The *Things to Do* calendar in my newspaper has been listing this event since the first of the month. I've literally waited weeks for this.

Vernon Jakobsen represents the Iowa chapter of the American Society of Dowsers and is here to discuss his craft. He is a man big in all directions—height, girth, and face. He sports a massive, scruffy gray beard that hides the top of the bib on his denim overalls and spans between the two shoulder straps. No mustache, though. With the right hat, if he wished, he could masquerade as a buggy-driving Amish man from nearby Kalona.

As he preps for his talk, some of the audience members take turns coming up front to shake his hand and say hello. Most attendees also made the trip from western Iowa and know him already and will stay for the Iowa Dowsers' monthly meeting to be held after the presentation.

Chairs are neatly aligned eight across and ten deep. Around two-thirds are occupied. It's 1990, so the only technology is an overhead projector for transparencies and a blackboard behind the tables. Vernon rolls the projector aside and centers himself at the front of the room, which goes silent.

"My name's Vernon Jakobsen. Call me 'Vern.'" His loud voice fills the space, especially resonant in the lower registers. "First off," he continues, *"dowsing, water witching,* and *divining* all mean the same thing: a tried-and-true method for locating hidden things."

He tells us that the most sought-after substance is water found in underground streams, and the most common tool is the dowsing rod. He favors L-shaped copper rods and scoops a pair off the table to demonstrate. Vern's meaty left hand, thumb side up, envelops a small length of shiny copper tubing. With his right hand, he inserts the short side of the L-rod down into it. The long end of the rod swings freely above his hand. After setting up a matching device in his right hand, Vern shows exaggeratedly how the rods swing left or right in parallel when a target is off to the side. When they're directly above the object, the rods cross, as if to signify "X marks the spot" on an imaginary treasure map.

"This one-hander's a beauty, too," he gushes while placing the L-rods back onto the table and picking up another tool. It's a metal stick, several feet long, bendy and bouncy. To simulate his rod finding its target, he holds it in front of him while he parades around the front of the room, the tool bobbing up and down absurdly. I softly giggle at what has to be an intentional phallus joke, but either it wasn't or nobody else gets it.

Next, Vern picks up a Y-shaped tree branch. Grasping the end of the fork in each hand, the single branch points out in front of him and bobs up and down with gusto, seemingly on its own.

The last dowsing tool is the pendulum. Vern explains, "It can be made of pretty much anything, but usually it's metal or crystal. It attaches to a string, or a light chain like mine. When it's above the target, the swing pattern changes. Dramatically. It'll go from back-and-forth to 'round in circles like this." He probably could have skipped showing us what "back-and-forth" versus "round in circles" mean, but the man is nothing if not thorough.

The presentation is sprinkled with folksy anecdotes about Vern's dowsing triumphs. Some of these are charming, emotion-filled stories—

how he found a beloved cat that had been injured and holed up for days under a neighbor's porch. Or how he was showing friends how the rods work, and they started wobbling strangely when near one of the "gals." That led her to make a doctor's appointment, resulting in an early cancer diagnosis that "no-doubt saved her life." The audience is mesmerized.

As Vern segues to a new topic, someone in my row chimes up.

"But does it *always* work?"

"It does for me!" Vern replies and further explains, "The key to dowsing is the person, not the device. It's almost like a psychic power. You think about what you want to find. The thing you're looking for activates the rods through your own mind and body. Anyone can learn it."

Vern softens his voice for an obvious humble-brag. "I'm still learning, but like a lot of people who've done this a while, I don't even need devices anymore. I just do this." He forms an "okay" sign with one hand, rubbing the tip of his thumb against the tip of his index finger. My fingers *tingle* when I'm standing near the thing I'm looking for."

To demonstrate the finger-tingle method, Vern holds his right hand straight out in front of him at shoulder height and forms the O. His eyes close for a few seconds as he gathers concentration. He opens them and embarks. "I'm checking for water pipes under the floor," he explains, meandering about between the presenter's table and the front row of chairs. Within seconds, he finds one running across the front of the room. He grabs the copper L-rods off the table behind him and walks through the same area. He is delighted when the rods verify the presence of the pipe. The audience is impressed and applauds.

"I said 'anyone can do this.' Can I have a volunteer?" he looks around the audience. A middle-aged man in a jacket and tie raises his hand. Vern motions him to the front and gives him dowsing rods. They soon locate the very same hidden pipe in the same place at the front of the room. His friends in the audience whoop with delight, and the rest of us clap politely.

Vern introduces one last claim about dowsing: "We could do this just as accurately with only a map."

Yes, you can dowse over a map without bothering to go to the physical location represented by said map. The sought-after substance will be found at the precise latitude and longitude indicated on the map by the dowsing tool.

Next topic: *ley lines.* Vern describes these as a web of straight lines of invisible energy encircling the Earth in many directions. They are not to be trifled with. Their power is multiplied wherever they cross, and they can be deadly at such locations. He warns that if your bed is at the intersection of ley lines, you should move it immediately to avoid brain damage. If your refrigerator is at an intersection, your food will rot faster and could become poisonous.

How could I have never heard of this until now?

Fortunately, dowsers will detect and neutralize ley lines for you. Vern re-scans the front of the room. This time, however, he moves his arms up and down, from knee-high to above his head, as though scanning for water pipes floating up in the air instead of underground. Instead of detecting the buried pipe he'd found a few minutes ago, he discovers a ley line crossing the front of the room at a height of around four feet. This also proves his earlier point that he only detects the things he's looking for.

Turning to the objects on the table, he reverently lifts up what looks like a piece of wooden front door art. It's a five-pointed star inside a hollow circle about a foot across. He calls it a *pentagram* and claims for it some very special properties.

"Pentagrams block lay lines," declares Vern. As evidence, he has a volunteer hold it by its edges so that it's centered on the just-discovered ley line. Sure enough, the dowsing rods no longer respond when moved through the area where the ley line used to be. "Oooooh," voice some in the crowd unsarcastically.

Vern announces that he'll do one final demonstration—*if* he can find intersecting ley lines. He starts tracking around the room. Traversing the back, he turns and heads toward the front along the right side of the seating area. The rods cross and stay that way. He's found a ley line running parallel to the side wall, at the same height as the other one spanning the front of the room. The lines converge near the front right corner of the seating area.

"Wonderful!" Vern declares.

Convenient! I say to myself.

Flouting the dangers he'd just described, he stands bravely at the invisible intersection. "We have ley lines crossing each other right . . . here!" His rods swing crazily. "Can I have another volunteer?" he asks as he returns front and center.

A woman raises her hand. He invites her up.

"Your name, ma'am?" Vern asks.

"Gertie" was her reply. She's a spry granny type, mid-seventies or so. I recognize her as one of the people who came up and chatted with Vern before the start of his presentation. She obviously trusts that Vern will protect her somehow from the harmful ley lines.

Vern stands facing Gertie, looming tall above her. He forms the okay sign with each of his hands, then interlaces the Os in front of his chest.

"Okay, Miss Gertie. I'd like you to grab my wrists and try to pull my hands apart."

Her petite hands are just big enough to grip on the narrowest parts of his wrists. She holds tight and makes several hard tugs. The Os stay interlaced. No surprise given the sheer bulk of his muscular arms.

"Now, let's try something . . ." Vern says. He escorts Gertie over to the side and parks himself at the ley line intersection. "Please try again now."

Her first tug pulls apart his hands with ease. Gasps and murmurs rise from the audience.

Vern exclaims, "Wow, I was really trying not to let you do that!" Gertie returns to her seat, grinning ear to ear, the conquering heroine graciously accepting her applause with a shy wave.

At this point, we're an hour into the presentation. Vern calls for us to take a short break, after which he'll open it up to a Q&A.

An urge to interrupt with some skeptically loaded questions has been building inside me. I've stayed quiet because at my feet is a backpack containing a surprise I've prepared in advance. My plan is to reveal it to the whole room after the break. I've never tried anything like this. It will involve making a bit of a spectacle of myself. And it will give Vern a chance to prove dowsing is legit.

I'm sweating like a sinner in church.

After the break, Vern stalls my plan by adding another demonstration. He asks Gertie if she'd come up for another kind of test. She agrees. Vern explains to the group what he wants to do.

"Just in case you're not convinced, I'm gonna switch places with Miss Gertie here."

Vern moves her to the spot away from the ley lines where he stood earlier. He instructs her to make the double O signs, just like he'd done before, and to link the Os.

"Hold on tight!" he tells her.

He reaches down, grasps her frail, meatless forearms, and gives them a little tug. Nothing happens. He spreads his feet a bit, squares himself up, and tries again with a grunt and a grimace. Again, nothing.

"Can't do it!" Vern declares to the audience's delight. "Now, let's move over here."

He brings Gertie over to the ley line intersection and asks her to link her fingers as before. He takes hold of her forearms. Her strength is now sapped by the ley lines, so he easily separates her hands with one little tug.

I can't take any more of this.

I raise my hand before Vern can send Gertie back to her seat. He calls on me and I ask, "Could I try pulling her hands apart over there?" motioning to the spot where he'd just failed. He looks at Gertie, and she nods with consent.

"Sure!" he says, waving me on up.

Suddenly, I'm very nervous. I may be the lone skeptic in the room, but I don't want to be seen as the enemy. I also don't want to appear foolish should Gertie be stronger than she looks.

I come face-to-face with her. She looks even older and frailer up close. Without instruction, she assumes the linked-Os position. I can see the whole audience focused hard on what I'm doing. I dry my palms on the thighs of my jeans and gently take hold of Gertie's forearms. I pull her hands apart with one easy tug.

She yelps "Ouch!" loud enough for all to hear and throws a hard glare at me.

The thing is, I really didn't pull very hard. It was nowhere near as forceful as Vern made his failed attempt appear to be. But with her one-word exclamation, Gertie makes it look like I used excessive force. In the eyes of the crowd, she probably invalidated my effort.

"I'm so sorry!" I tell her, realizing too late that I may have just validated her invalidation.

True believers in dowsing won't be swayed by the bad man who hurt Miss Gertie. I didn't expect to convince them. It's the fence-sitters I hope to reach, the ones unsure whether or not they buy into all of Vern's claims. At this point, I've probably not done much to move anyone into my camp. But that's okay. None of this business with the ley lines has anything to do with what's in my backpack.

Vern sends us back to our seats and opens up the Q&A. I'm the first to raise a hand. He calls on me as though unbothered by what I'd just done to poor Gertie.

"Could you try one more thing?" I ask respectfully but vaguely.

"What would you like me to do?" replies Vern.

Not knowing for sure whether I'd have the guts to intercede on behalf of Science and Reason, I'd packed some materials for a simple experiment. I'll have to be careful not to make Vern feel threatened.

"You said that you can find anything you want with the dowsing rods, right? Just by thinking about it?"

"That's right."

"How about copper?" I suggest.

"Sure."

"Does it have to be buried underground?"

"Nope."

The moment of truth. I dip into my backpack and pull out a red plastic beer cup about three-fourths full of pennies. I get up and boldly head to the front of the room, not waiting to be invited. As I'm walking, I say, "Great! So dowsing rods will detect this cup of pennies?"

"Of course," he replies, reaching over to the table to fetch his L-rods.

"Okay then!" I put the cup on the floor. He approaches it and the rods cross directly above it. Again, no concerns about possible interference from the buried water pipe or the ley lines.

"Anyone else want to try?" he asks the audience.

A teenage girl volunteers. Perfect timing. While she's coming to the front, I scurry back to my chair, grab my backpack, and return to watch the girl dowse for the cup of pennies. She succeeds on each of several passes. The audience loves it.

I pipe up: "Would it matter if I put this card over the cup?" It's a standard four-inch-by-six-inch index card, blank on both sides. Vern and the girl look at each other, acknowledging wordlessly that the card won't matter. They probably reasoned, as I did, that if you can dowse for something hundreds of feet below the ground, a thin paper card shouldn't get in the way.

It doesn't. They each test the rods on the card-covered cup of pennies, and both succeed. More applause. Thankfully, the audience is really into this now. I'm glad they're not hostile, but their raptness makes me even more nervous about what's coming.

The girl returns to her seat. Vern turns to me and asks, "Anything else?"

Once again, so all can hear, I say to Vern: "Maybe. Now would you like to *really prove* dowsing works, just in case there are any skeptics left in the room? If you succeed, I'll pay your expenses for coming here today." I reach into my pocket and pull out five twenty-dollar bills. "If you don't want it, you can donate it to the Iowa Dowsers, or to charity." I put my money on the table.

"What do you want me to do?" he asked.

"Simple . . ."

I take two empty red cups from my pack, and two more blank index cards, and put them on the table a few feet apart. I also pick up the matching cup from the floor with the pennies in it and line it up with the others on the table. I cover each with a card, trying not to let on that my hands are trembling.

I explain to Vern, "All you have to do is dowse for the pennies—*just like you did a minute ago.* I'll give you ten tries. After each one, you turn your back and I'll shuffle the cups around. You get a hundred dollars if you find the pennies seven or more times. Easy, right?"

He stands there for a moment and looks at the set-up on the table, under the gaze of a hundred expectant eyes. Maybe he's thinking of the words from the American Society of Dowsers homepage:[2]

We must accept dowsing as fact. It is useless to work experiments to prove its existence. It exists. What is needed is its development.

He shakes his head no. There's disappointment in the audience, for they're sure he'd have met the challenge.

But Vern doesn't merely decline the test. He becomes outraged. At me. His face reddens, he points a dramatic finger at me, and bellows, "***Jesus rebuked the skeptics!***"

Now all those eyes turn to me.

Jesus did do lots of rebuking in the New Testament ("Oh ye of little faith . . . ," Matthew 8:26), but Vern isn't Jesus, he's not here to affirm his identity as a savior, and my questioning his claims isn't sacrilege. He's here to demonstrate that dowsing works. My little test is the only chance today to see if that's really true.

All I say in reply is, "Are you *sure* you don't want to show that it actually works?"

"I said 'No,'" he replies. "It works fine. Let's take some questions."

BACKSTAGE

It's possible that Vern wasn't so much worried about failing a test as he was thrown off by the weirdness of my proposal. He's probably only ever demonstrated his dowsing ability in one of three ways. First, by finding water in the field. More on that later. Second, by dowsing for something that's right there, plainly evident, like the cup of pennies. Third, by finding something that can't be verified because it's invisible, like ley lines, or too costly, like buried pipes. Unfortunately, none of these is a real test.

Despite Vern's stern talking-to, I was okay with the way things ended. I got the chance to show him, and a potentially hostile crowd, what an actual test could look like. I was respectful, even when Vern didn't reciprocate. There was also a small personal triumph: In front of a roomful of strangers, I overcame my anxiety and called out a self-proclaimed expert who was issuing bad information. Maybe there were even a few in the audience who got the point of my proposed demonstration.

Why did Vern all but accuse me of putting the "demon" in "demonstration"? The truth is, I only asked him to do what he repeatedly and publicly claimed that he could. The tabletop experiment would have been simple and fair.

As for the test itself, I should say a bit more about why experiments are the gold standard for validating claims.

First, they clearly define success and failure. As with most dowsers, Vern claimed a very high success rate. A test requiring as few as seven "hits" in ten tries should have been a low bar for him. On the other hand, if he really couldn't dowse, his chance of getting seven or more correct by blind guessing was only 2 percent—a risk I was willing to take with my hundred dollars.[3]

Second, *informed consent* is an important ethical concern in experiments with human subjects. I made sure to explain to Vern the test conditions and success criteria and asked for his explicit agreement to participate.

Third, all proper experiments use *randomization*. I'd planned to randomly shuffle the three cups for each of ten trials.[4] More elaborate field experiments use the same principle, under conditions agreed upon by the dowsers. For instance, a study might have a single pipe buried underground, with water running through it, or not, in a random sequence of trials. Or there may be several buried pipes with water running through one of them, randomly selected on each trial. Dowsers have never demonstrated that they could either detect water or track its path beyond chance levels in such experiments.[5,6,7]

Fourth, larger sample sizes are better. This could mean testing many dowsers. Or it could mean testing a few dowsers using a large number of trials. Having only Vern and only ten trials is far less than I'd have preferred. But it seemed a reasonable compromise for a field test, considering that more trials would have bored the audience and been even more of a deterrent for Vern.

Fifth, *double-blinding* is one of the most powerful experimental tools, but too often, its praises go unsung. Its purpose is to prevent subjects from picking up clues that might bias their responses. The cup test was *single-blind* because only the subject, Vern, would have been in the dark as to the correct choice. A double-blind test would go further by blinding Vern, me, and anyone else in contact with Vern to the correct answer—including observers in the audience who might subtly cue correct responses in ways Vern could have detected. Logistically, this would have required another assistant, plus more time that I couldn't assume I'd have.

Finally, it's a good idea to eliminate in advance any excuses the dowsers might have about failing to perform well. This means, for instance, having them first demonstrate that they can detect the targeted substance and verify there is no interference from other substances at the test site. Vern performed this check when he detected the pennies on his first try, unimpeded by pipes, ley lines, index cards, common sense, or anything else.

How fantastic it would be if tests revealed that dowsing works! Lost objects and rare substances would be found with the wave of a wand. Water for agriculture, archaeological relics, buried treasures, underground utilities, and the remains of missing persons would be hidden no more. Dowsers around the world could celebrate, unfettered by Doubting Thomases or Skeptical Barrys. Science would devote massive attention and resources to understanding the physics of this new type of natural force.

The evidence suggests otherwise. As I noted earlier, dowsing has been tested in a number of rigorous, well-designed experiments. They've been run with the cooperation of dowsing organizations, used substantial numbers of dowsers across large numbers of trials, and implemented randomization and double-blinding protocols. In all of these tests, dowsers have never demonstrated a better-than-chance ability to find anything.

There's another good reason to be skeptical of dowsing that rarely gets discussed: How could it possibly work? How could centuries of research in fields such as physics, geology, and materials science have failed to explain such a supposedly robust phenomenon?

Dowsers offer several explanations. The most common is that their dowsing tools respond to targeted substances the way that magnets attract iron. That sounds reasonable, but it's a false analogy. The science of magnetic forces is well-established. They're restricted to only certain materials within relatively close distances. In contrast, literally nothing is known about dowsing forces. Long-accumulated knowledge in science finds nothing that makes water, metal, or paper maps interact from a distance with metallic, wooden, or crystal dowsing implements or with human brains. Nothing's impossible in principle, but it's important to know that *if* dowsing is real, it would have to be based on a potent force operating right under our noses that has never been identified, isolated, or measured.

Not knowing how it works is a problem that should be a mighty concern for dowsers. It isn't. They simply don't care.

It should also concern dowsers that everything about their craft can be explained without invoking any unknown forces. For instance, when searching for water on a client's property, a dowser typically tells the owner precisely where drilling a well will tap into an underground stream. For this service, they do, in fact, have a very high success rate of finding water. But the reason for the success has nothing to do with dowsing. Underground streams are rare. What actually happens is that dowsers instruct landowners to drill into the water table—vast regions of water-saturated earth underlying the surface of much of our planet.[8] So when the dowser says to "drill here" and water is found, they almost always could have drilled anywhere in the area with equal success. The dowser's tools were not detecting anything, but nobody's going to check this by paying to drill a second hole at a control site fifty feet away.

How do we explain the movements of the dowsing tools—the swinging rods, the bobbing branch, the twirling pendulum? It's simply the dowser's own hand movements, the so-called *ideomotor effect*.[9] This is a well-documented, unconscious muscular response to suggestions and expectations.[10] Dowsing rods and pendulums are inherently unstable, sensitive to tiny movements of the hands. That dowsers are unaware of these movements makes it feel as though the rods are responding to an outside force. Support a dowsing tool in its own stand or bracket, untouched by human hands, and it will refuse to cooperate no matter what substance or object it's hovering over.

Last, I want to address the question of whether there's any danger to dowsing. It seems innocent enough, even if it doesn't really work. But a fuller answer to the question is that it's indeed dangerous—in proportion to the stakes attached to accurate detection. If you're dowsing for water pipes in your backyard prior to excavating for a swimming pool, the worst-case scenario is that you fail to detect a pipe, it breaks during the dig, and you get it fixed.

But when utility companies dowse for buried water and sewer lines,[11] they're wasting time and incurring costs that get passed along to consumers. When cash-strapped school districts buy dowsing devices for detecting drugs in student lockers, they're spending scarce resources on a fool's errand.[12] Perhaps most heinous of all, bomb-detecting dowsing rods sold to national governments at high prices have failed to detect explosive devices which detonated and cost hundreds of lives.[13,14]

Several people approached me after Vern's presentation, though far fewer than the number that swarmed him the moment he finished. But those who came to me were all appreciative for a skeptical angle. One young man around eighteen years old stands out in my memory. He was the first to seek me out.

"That was really awesome what you did," he said. "I wondered if there was anything to dowsing, and now I think it's all just a bunch of BULLSHIT! You made me a skeptic."

I replied, smiling, "Maybe it's enough just to say the evidence is poor and leave it at that."

"Yeah," he said. "I get that." And then he added with complete earnestness, "But have you looked into alien abductions? There's tons of evidence for *those*!"

CHAPTER TWELVE

The Yogi Has No Robes

Come on Inner Peace! I Don't Have All Day!
—Book by Sachin Garg[1]

FRONTSTAGE
It feels good to do good. Big things, like raising a kind and happy child. Or little things, like buying a cup of coffee for a stranger. Good deeds feel good because something is given up—usually, time, money, effort, or some combination—purely for the benefit of others. The greater the sacrifice, the greater the good.

What if you could do good on a global scale, immediately, directly, cheaply, and effortlessly, by sitting quietly with some other people for a short time each day? And what if, as a bonus, you'd also achieve perfect health and develop superpowers, such as unaided flight and invisibility?

Millions of people not only subscribe to these claims but also believe science fully validates their effortless philanthropy.

It's the early 1980s. I've developed a benign addiction to research on extraordinary phenomena. It's only a side-interest to my real work, but it's a fun diversion. I've discovered a steady stream of extraordinary reports in peer-reviewed journals claiming support for extra-sensory perception (ESP), alien visitors, faith healing, reincarnation, astrology, and more. Definitive proof seems right around the corner.

But with each passing year, we're no closer to the corner. Proof hasn't materialized for any of these things, and I'm tired of all the low-quality videos of ghosts and UFOs; of ESP demonstrations that won't replicate;

and of poorly designed faith healing studies designed and conducted by biased zealots. I'm especially repulsed by charlatans who bolster their egos and bank accounts by hard-selling phony cure-alls to the physically and psychologically vulnerable.

I miss the stories that once raised the little hairs on the back of my neck and ignited my paranormal imagination.

One day, I'm having lunch at my office desk, listening to the news on the radio. Just before a commercial break, there's a teaser for an upcoming report about a presentation at the Iowa Academy of Science conference:

> Does meditating really improve the weather? More, after this word from our sponsor, Lunker Land!

I'm thinking, *This must be the TM people. Should be interesting.* And of course, I'm also thinking *What the hell is "Lunker Land"?*

Lunker Land is a set of small man-made ponds near Cedar Rapids, explains the over-caffeinated announcer. They stock these waters to the gills with hungry bass, and you pay a fee to drop your line and take your chances. Soon, you, your toddler, or a fencepost can catch a string full of "lunkers" to take home.

As for the meditation story, it is about TM, and it's *very* interesting.

TM is Transcendental Meditation, a practice rooted in the teachings and preachings of Maharishi Mahesh Yogi. "TM" also refers to the organization that promotes and supports the practice.

Beginning in the early 1960s and on the heels of a relatively quiet post–Korean War era, the United States experienced major social upheavals. Civil rights demonstrations, the women's liberation movement, anti-war protests, and environmentalism all gained momentum. It was scary and loud, reflected in rock and roll music's defiant soundtrack for the times.

Against this backdrop, TM offered a cure-all. It vanquishes stress, they claimed, and improves alertness, sleep, and cholesterol levels, while also relieving depression, anxiety, hypertension, addictions, and pain.[2] If enough people adopt the practice, poverty, crime, and social conflicts will disappear.

The Yogi's fame rose to new heights after 1968 when the most famous rock band in history stayed for a time at his compound in

India. The Beatles' endorsement piqued widespread interest. TM centers popped up around the world, along with an extensive teacher-training network. The movement flourished.[3]

Ads for introductory TM courses were commonly posted around my college campus. In addition to reducing stress, they claimed you'd develop special skills, such as enhanced senses, levitation, and invisibility. I was skeptical, since no other extracurricular group offered superpowers—at least not without having to die first.

Fast-forward. It's the first month of my first faculty job at the University of Iowa. I heard that TM founded Maharishi International University (MIU) just an hour away at the defunct Parsons College campus in Fairfield. MIU's several hundred students live and breathe TM while studying the guru's philosophy. They're told they are creating "Heaven on Earth."[4]

Like most non-meditators, I have doubts—especially recalling their ads for levitation and invisibility training. Nothing against meditation as a relaxation technique, but TM's exaggerated claims and cringy evangelical confidence make me skeptical. But I'm still curious.

One spring Saturday, I drive the hour down to MIU to see the campus for myself. Four lanes narrow to two amid the rolling cornfields, and soon I'm in tiny Fairfield.

MIU looks like any other small college but with two colossal exceptions: a pair of round, low-slung, gold-domed meditation buildings, each nearly two hundred feet across. A *New York Times* reporter described them as "gilded breasts,"[5] but to me, they look like flying saucers grounded for repair.

I approach the "Men Only" dome and glance around to see if anyone might object to my entering. I see no one. The entry door's unlocked so I step warily into the deserted, cavernous space. "Anyone home?" I call out but hear only echoes. There's paraphernalia strewn about—pads, mats, books, gongs, pillows—and a funk in the air that I can't identify and don't really want to.

Exiting the men's dome, I head over to the women's a few hundred feet away. Respecting MIU's gender segregation rules, I only peer inside through the glass door. It's virtually identical to the first dome and just as deserted.[6] Maybe all the meditators are doing the invisibility thing.

Strolling around campus, I soak up the warm sun and positive vibes. With administrative offices and classroom buildings shuttered for the weekend, only a few scattered pairs and trios of students amble about. Most say hi and smile when our paths cross.

After an hour or so, I grab some pamphlets at a kiosk, return to my car, and drive back to reality.

The building of the golden domes was a major project for MIU. The schedule was such that their concrete foundations needed to be poured during the frigid winter months. But pouring required above-freezing temperatures. What could the MIU community do to avoid costly delays?

To tip the weather odds in their favor, MIU students meditated as a group on evenings before scheduled concrete pours. An atmospheric physicist on the MIU faculty, Dr. Robert Rabinoff, later went back and noted the next day's temperatures. Nearly all were above freezing. He presented his findings to the Iowa Academy of Science, *crediting TM for causing the warmer weather*.[7] If he's right, the implications would be profound—not just for construction projects but also for physics and agriculture.

For a PhD atmospheric physicist to make this claim lends it scientific legitimacy, and the story revives the prickly raised-neck-hair feeling I've so missed. I've seen no evidence for flying and invisibility, but maybe there's something to *this* claim.

While looking into Rabinoff's work, I find other unusual TM projects. MIU scientists are especially proud of one in particular, and it sends shivers up my skeptical spine.

The research was conducted in Jerusalem over a sixty-day period. In it is a graph that I find uniquely captivating.[8]

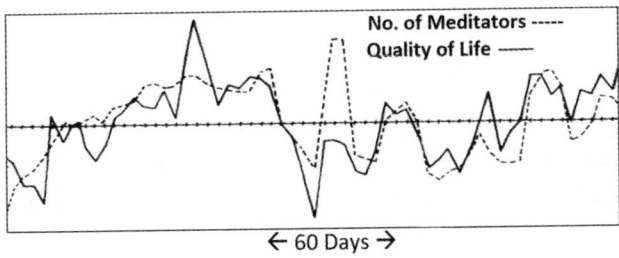

The dashed line represents the sizes of daily meditation groups meeting in August and September. They ranged from 65 to 240 people. The solid line is a Quality-of-Life Index that combines data on crime rates, fires, car accidents, national mood (inferred from newspaper headlines), stock market, and war fatalities. Higher values indicate higher quality of life.

The results are striking: Over time, the Quality-of-Life Index eerily tracks the meditation group size. I've never seen a real-world correlation this strong in any published study, and the researchers verify it with advanced statistical analyses.

The Jerusalem study is published with a theory asserting that group meditation alters the Unified Field of physics, the basis of all matter and energy. It claims that the group practice of TM injects coherence and harmony into the field, in turn triggering coherence and harmony within everyone it reaches.

If this so-called Maharishi Effect and its supporting theory are true, it would change everything. It would shred the fabric of science, painstakingly woven over centuries, and arguably the greatest of human achievements.[9]

Physics, neuroscience, and other fields would need to rebuild on new foundations.

We'd also have a blueprint for merging spiritual practices and hard-nosed science.

It would be proof of humankind's interconnectedness—not via bonds or genetic codes, but through our ties to the Unified Field.

Even if the Yogi is wrong in some of the details, wouldn't it be fantastic if there's really something to it?

Or is it just a pseudoscientific word salad, dressed in deceptive statistics?[10]

BACKSTAGE

Students of TM are taught they're on the path to "unity consciousness," a superior way of life and higher form of existence.[11] It's the axis around which MIU's curriculum turns, guiding the lives of every practitioner.

Most participants will tell you that doing what they're doing makes them feel fantastic.

TM, the organization, says the individual benefits are proven to exceed those of other meditation techniques. Their own research would appear to support this claim. Independent studies disagree.[12] However, my focus here is on an extraordinary claim that goes beyond the individual benefits of meditating. TM researchers claim that, when a group practices TM together, meditative effects extend outward to affect non-meditators. The effect is instantaneous and spans any distance. This phenomenon is called *superradiance*, better known as the Maharishi Effect.

Is this possible? It really all comes down to evidence. So let's take a closer look at the two studies mentioned in the Frontstage section.

Did TM Make Iowa Warmer?

Rabinoff claimed TM groups raised temperatures in wintry Fairfield, Iowa. If TM actually plugs meditators into the Unified Field, group meditation could theoretically affect the environment. So it may not be such a weird idea after all, and I'm not qualified to rule out the possibility.

I am qualified to judge Rabinoff's study, however. It's true he observed above-freezing temperatures on 75 percent of the days concrete was poured. Not bad. But 72 percent of the ninety-two days in the relevant period had above-freezing temperatures anyway. Even aside from that issue, there were only eight attempted concrete pours to work with. That's simply way too small a sample to conclude anything.

An even bigger issue is theoretical: the string of assumptions hidden behind Rabinoff's claim. It goes something like this: Meditators are assumed (1) to produce a kind of energy, which (2) combines with other meditators' energies, (3) gets amplified, (4) spreads to the environment, (5) moves around air molecules in very specific ways, and (6) raises the *next* day's temperature. Every step is dubious and, given the extraordinariness of the claim, should have been verified with appropriate measures.[13] Rabinoff didn't verify any of them.

Things got even worse for the study when a local scientist threw cold water on the warm weather claim.

Franklin Trumpy from the Des Moines Area Community College looked into Rabinoff's claim. He learned MIU was its own general contractor for the dome construction project. As such, they'd request concrete from a local company for the foundations on an as-needed basis.

Trumpy next contacted the concrete company and asked about their procedure. They told him their procedure is to decide the day before a planned pour whether or not to follow through with it the next day. So if they plan to come, they inform the client the day before.

How did the company decide when to pour? By checking the next day's forecast from the National Weather Service. If the forecast was above freezing, they'd tell MIU they're pouring tomorrow, and MIU held a group meditation. If the forecast was below freezing, they'd cancel the pour, and there'd be no group meditation.

This means Rabinoff had it exactly backward: Meditating didn't cause a change in the weather; the weather caused a change in the meditation.

Six years after Trumpy published his critique,[14] Rabinoff published his original conference presentation, *unchanged*, in a TM journal.[15] He must have known his study had fatal flaws, but published it anyway. MIU also continues to list it at their site as evidence of the Maharishi Effect.[16]

Recently, I emailed Rabinoff, asking:

Did you ever follow-up your paper about TM affecting weather in Fairfield, such as either retracting it as presented, or expanding the number of trials to make it stronger?

He replied the same day:

The short answer to your question is no.

The longer answer was still rather short, explaining that it would have been difficult to do follow-up studies and that his main focus was on teaching.

Compared to the media attention paid to Rabinoff's flawed research, Trumpy's critique got little notice. In the hallowed tradition of "Man Bites Dog" journalism, you get more publicity saying you can change the weather than saying you can't.

Did the Jerusalem Study Improve Quality of Life?

When I read about the Jerusalem study and saw its uncanny graph, it affected me. I wondered: *Could the Maharishi Effect actually be for real? Could a relatively small group of people practicing TM alter the brains and behaviors of millions of other people?*

Most TM research doesn't make such extraordinary claims. For example, they've found that, like napping, meditating can lower blood pressure.[17,18] But the Maharishi Effect is different. Its central claim is not accepted by mainstream science, despite TM promoting it as a proven scientific breakthrough.

The Yogi proclaimed that if a group of advanced "Siddhi" meditators number at least the square root of 1 percent of a population, it benefits everyone in that population. A group of one hundred meditators would impact a million other people. A group of roughly nine thousand would be enough to heal, enliven, and unstress everyone on the planet.[19]

If that's true, science would have to reformulate bedrock assumptions about the nature of the physical universe. But before throwing out a few thousand years of scientific progress, we first must consider the possibility that the research is flawed, that the researchers were fooling themselves, or that they may have purposely misled the public for their own gain. These problems are not as rare in science as we'd wish.[20]

Trumpy showed how Rabinoff appeared so convinced the weather synchronized to group meditation that he didn't weigh any other possibilities. Did the Jerusalem study also neglect factors that caused the daily variations in Quality-of-Life Index and TM group size to sync up?[21]

TM researchers on the Jerusalem study weren't naïve. They knew the main correlation they tested could have been enhanced by factors other than group meditation. So they statistically controlled for things that could affect both the population's quality of life *and* the meditators' decisions whether to join the day's meditation group. For example, they statistically controlled for the possible effects of weekends and two Jewish holidays. Either factor could affect quality-of-life measures (people are happier on holidays and weekends) *and* provide more free time to participate in the group activity.

But crucially, the study didn't control for many other factors. This is its biggest flaw, yet it's subtle because it involves seeing something that's missing from the picture.

TM practitioners believe their collective actions foster "Heaven on Earth." TMers living in or visiting the Jerusalem area undoubtedly were eager to do their part for the study. It would have been reasonable for them to monitor the daily news and show up to meditate when they felt most needed.

To illustrate why this would be a problem, suppose Jerusalem's Monday morning news reports an uptick in weekend casualties in the nearby Lebanon War. This would likely cause more meditators to attend the Monday evening meditation group. Flare-ups being temporary, this one peters out Monday all on its own and completely dissipates by Tuesday. But the researchers' models would be prone to infer that Monday evening's meditators *caused* Tuesday's calm.

Could there be other mundane causes for the correlation in the graph? My Iowa colleague, Dr. Evan Fales, and I looked into this.[22] We found a number of contemporaneous events that likely would have biased the Jerusalem results. A few examples:

- Three additional Jewish holidays
- August was vacation month for most Israelis
- Israeli Prime Minister publicly announced his plans to resign
- Prime Minister resigned
- Major battles for control of Beirut
- The Pope calls for an end to the war
- The United States announced escalated military involvement

TM researchers controlled for *some* cultural and environmental factors but not for others. This is a big clue that there may be problems with the study.

Did the TM researchers curate their measures to get the results they wanted? Possibly. There's a famous adage among statistics nerds: If you torture the data long enough, it will confess to anything.[23] TM researchers took five years to publish their results. Moreover, they used a research design that lends itself to abuse and bias. They also had a problem just like the weather study's: The extraordinary assumptions needed for the effect to happen weren't verified with more direct measures. In short, ultimately this evidence is unconvincing.

The Data-Sharing Saga
Researchers are bound by professional codes of ethics to share data from their published research. I asked the TM researchers numerous times for a copy of their Jerusalem data set, but it was to no avail. Doing so could have vanquished my suspicion they'd made some serious mistakes, whether inadvertent or intentional, that caused the results to look more compelling. I would have rerun their analyses, verified their findings, and retested their hypotheses, controlling for some of those factors they omitted. None of this would have suppressed the Maharishi Effect unless, of course, it never existed to begin with.

Why were they so reluctant to share the data? I'll never know for sure. But any scientist would be skeptical of a study whose authors shield it from scrutiny.[24]

I probably annoyed lead researcher David Orme-Johnson by requesting the Jerusalem data so many times. I wrote him a series of snail-mailed letters outlining my intentions. He responded with a series of bogus excuses: that I was unqualified; made false claims about their work; had a cultural bias; wasn't trustworthy; they'd already shared it with others; they paid for the study, so didn't have to share; and so on. He also made weird stipulations if he were to share the data, such as demanding to insert whatever material he wished into any report I wrote up.[25]

Early in my pleas for the data set, Orme-Johnson insisted that he had shared it with two other non-TM researchers. But he refused to identify them or direct me to a written report. I suspect the researchers don't exist. But if they do, and they confirmed the Jerusalem results, their findings would have been trumpeted at TM websites.

Could the Study Be Improved?
Stopping short of accusing the researchers of fraud or willful deceit, political scientist Philip Schrodt faulted them for "basic measurement

errors and a poor understanding of experimental design which led to statistical self-deception."[26]

One of my statistically inclined nonexperimentalist colleagues once noted that statistics are a complicated and weak substitute for a good experiment. The statistical methods used for the Jerusalem study were indeed very complicated. A well-designed experiment would have been much simpler, less expensive, and yielded more definitive results.

Here are all the changes that would have been required:

1. Randomly assign the number of meditators who'll participate in each day's group.

That's the whole list.[27] Randomizing practically eliminates the chances of bias from news events, personal motivations, holidays, and so on. But the TM researchers always opt for weaker, more complicated, less reliable approaches. Why is that?

I asked them directly, and they responded by informing me that randomizing the group sizes would be immoral. It would have meant artificially reducing the number of potential meditators on some days. They claimed this would cause people to die who'd otherwise be saved. I pointed out that scientists testing potentially life-saving drugs apply randomization, regardless of their confidence in the outcomes. This is how you rule out biases and excluded factors. But the TM researchers argued that it doesn't matter what others do. They had to abide by their own ethical code.

I also told them they'd probably save many more lives in the long run by using stronger methods, generating convincing results, and having government agencies such as the National Institutes of Health support permanent TM groups. No dice. A moral code is a moral code.

Anyone looking into the TM research program would quickly see that the Maharishi Effect studies are conducted largely, if not exclusively, by advocates, practitioners, MIU faculty, and others affiliated with TM. This doesn't automatically disqualify the work, but it's a big concern. Biases, like cockroaches in the best of homes, can sneak in through the cracks. TM researchers should be eager to dispel any whiff of bias. It would help to collaborate with qualified scientists who have no personal stake in the results.[28]

Maybe a randomized experiment without life-and-death ramifications would be more to their liking. I offered to collaborate with the Jerusalem study authors on a simpler, tighter, smaller-scale Maharishi Effect study. They'd already claimed that people practicing TM in one room of a building will impact the brainwaves of non-meditators in another room.[29] But the sheer volume and complexity of these data open them to multiple interpretations, depending how they're analyzed. Why use such a complex and indirect approach when a simpler, more direct method would be easier to implement?

I proposed randomizing the meditation group's schedule for meditating. People in another room would work on a set of written problems, unaware of the meditation group's activity. If the Maharishi Effect is real, the non-meditators should score higher when meditation is happening than when it isn't.

I stipulated that we run half the subjects in each of our respective labs. I offered to cover all the costs—theirs and mine—with my own research funds.

The TM researchers took my offer under advisement but declined it several weeks later. Their only explanation was that they'd already proven the Maharishi Effect to their satisfaction.

In my opinion, TM researchers avoid more definitive research designs because they would definitively falsify the Maharishi Effect claim. They would not like this outcome, to put it mildly. The difference between us is that I didn't have any skin in the game and would have readily admitted it if I were proven wrong.

Is the Maharishi Effect Research Ethical?

Maharishi Effect research seems less strange if you think of TM as primarily a business institution that uses science in an attempt to legitimize its products and services. But the intrusion of business into scientific research raises ethical concerns and is known to compromise research integrity.[30] Ideally, research should be unfettered by financial interest in the results. TM pays for much of its own research and has a vested interest in the results.

They also exalt the research as if it were established science, using it to sell their meditation courses, Maharishi Ayurveda health and wellness supplements, Maharishi Astro astrology paraphernalia, Maharishi Vedic City homes, TM conferences, group retreats, and a political wing called the Natural Law Party.

Despite all the activity, a half-century of research on the Maharishi Effect has had no discernible impact on physics, physiology, psychology, sociology, or any other field.

Like less commercialized forms of meditation, TM works by triggering the *relaxation response* in the brain and body.[31] Most practitioners find it delightful and beneficial. In contrast, only the TM organization appears to benefit from the Maharishi Effect—via waves of cash rather than unity consciousness. By blurring the line between science and commerce, they thrive today despite a long record of minimal impact on science.[32]

I'm not a business ethicist, but I can say with some authority that abiding by standard codes of research ethics was never a priority for the Maharishi Effect researchers. A conversation I had over lunch with two of the Jerusalem study's lead researchers illustrates this nicely.

"What if I don't want to experience the Maharishi Effect?" I asked them. As psychologists, they must have been trained in the ethical treatment of human subjects. "Like if I'm an opinion writer and love to work when I'm emotionally charged and angry. Then the Maharishi Effect would be stripping me of my joy and possibly my livelihood."

"No," was the quick reply. "It'll only improve your writing."

Flabbergasted, I asked, "So there's no way I can opt out?"

"Why would you want to?" was the reply.

They're basically saying, "If we're successful, TM will be administered to everyone, everywhere, all the time, whether they want it or not. But don't worry. We know what's best for everybody."

No, sir, I don't believe you do. In science, nothing gets administered to whole populations without extensive pretesting, community consent, public health clearances, and so on. It's a complicated process, but it's there for good reasons.[33]

"What about informed consent?" I asked, switching to a related ethical issue. "You say the Maharishi Effect literally alters the minds and brains of everyone in its path, right?"

"Basically, yeah."

"But you administer it without anyone's approval—no institutional review board, no community officials, no public health department, not even the subjects themselves. Is that even ethical—bypassing formal review and informed consent?"

"It would only be unethical if there were any risks," one of them replied, "and there aren't any. We *know* TM does nothing but good. Zero potential for harm."

"But it's really not your place to make that call." I thought I'd just dealt a death blow to their argument, but they shrugged it off.

They can't "know" TM is risk-free without willfully ignoring the evidence that TM *can* harm some people, for example, by exacerbating preexisting psychological problems.[34] But that aside, if I wanted to conduct a field experiment on how personality affects ice cream preferences, I'd have to satisfy a slew of criteria designed to prevent harm and abuse.

First, I'd need approval for the research protocol and use of human subjects from my university's Institutional Review Board.

Second, my subjects would have to be volunteers who could opt out at any time. I can't force-feed ice cream to everyone in a community or region, even if I have enough to go around and personally believe it to be harmless.

Third, I'd need every subject's written consent to participate.

Fourth, I'd have to inform every subject of the potential risks, including brain freeze and the potentially embarrassing consequences of lactose intolerance.

Fifth, I'd have to give subjects contact information to report any questions, concerns, complaints, or adverse effects. I'd surely be banned from doing research if I were telling subjects "You may not *want* the ice cream, but you *must eat* the ice cream because I know what's best for you."

The Maharishi Effect research flaunts the usual safeguards. It's not that I think the Maharishi Effect exists and could endanger people.

A closer look at it made me seriously doubt the effect is real. But the researchers' practices say a lot about their more-enlightened-than-thou attitude and lack of any concern about potential harm.

Here's another thing I asked them: "With all the meditation happening in Fairfield, why are your crime rates so high relative to other places?"

It's true. After decades of group meditation, sweet little Fairfield (population 10,290) should be awash in chillness. Instead, their crime rate is in the state's top 6 percent.[35]

"But *some* crime rates have come down!" was the reply. "Sure, others are tougher, but we'll get there." Far from proving the Maharishi Effect, this shows that even psychologists can don the rose-colored glasses of confirmation bias.

The Jerusalem study is promoted by TM as if it were settled science, a shimmering jewel in its research crown. It really does look impressive to the untrained eye. But its problems are numerous and subtle. A physicist at the University of Iowa told me, "It's pseudoscience. Sophisticated pseudoscience, but still pseudoscience."[36] Another physicist, Robert L. Park, was less restrained and deemed the Maharishi Effect "a clinic in data manipulation" and "voodoo science."[37]

The TM researchers I've met were all very smart and well-trained. But even smart and well-trained scientists aren't immune from self-deception—especially when they excuse themselves from accepted practices such as data-sharing and responding to informed criticism.

In his book *Why People Believe Weird Things*, in a chapter aptly titled "Why Smart People Believe Weird Things,"[38] Michael Shermer argues that smart people are highly skilled at defending beliefs they arrived at for non-smart reasons. If TM researchers learned, loved, and identified with TM prior to their professional training, they may have later acquired rationalizations for their feelings, beliefs, and research practices through confirmation bias or other psychological routes.

Astrophysicist and science popularizer Neil DeGrasse Tyson once remarked in an interview:[39]

For me, science is a vaccine against charlatans. It enables you to know when someone is full of shit and when they're not. And that's power. It's power of protection for yourself so you will not be exploited by those who do not have your best interest in mind.

Science requires researchers to submit to safeguards against self-deception before their work gets taken seriously—especially if unorthodox claims are involved. To the safeguards I've already discussed, let's add humility. You should *want* to make your work better. *Seek* informed criticism. *Fix* weaknesses. These kinds of attitudes and practices only strengthen the work in the long run.

If you're a proponent of an unorthodox claim, you absolutely deserve to take your best shot with it. But you also need to heed those who take the time to offer informed criticisms and corrections. To keep "replicating" the Maharishi Effect—as TM claims it has done over fifty times—without fixing design flaws will never build its legitimacy in science.[40] It will only continue to erode it.

Constructive skepticism plays a critical role in keeping research honest. Skeptics identify bad practices and remind whoever's interested that claims which fly—or turn invisible—in the face of established science must be held to higher than usual standards.

CHAPTER THIRTEEN

The Power of the Pyramid

"Pyramid Power" is the biggest story we should never have done.
—Adam Savage[1]

Frontstage

Shane's a small-town Iowa boy attending the state's mega-sized flagship university. First in his family to go to college, a little out of his depth as a freshman and sophomore, he's now a seasoned junior snarfing up classes like Tater Tot casseroles and rhubarb pies.

One of those classes is Sociology of the Paranormal, which I'm offering for the first time after years of thinking about it. I'm recently tenured at the University of Iowa. Reveling in my academic freedom, I'd opted to teach something a little out of the ordinary. "A little" is an understatement. No other sociology department in the country offers it. Naturally, some colleagues are skeptical about this new course on skepticism.

Shane's only previous experience with sociology was the requisite Intro course 101. He's using mine to fulfill a social science requirement. He's also taking a physics class to knock out a *real* science requirement. For reasons known only to that professor, Shane and his classmates were assigned to read a short, uncritical article on pyramid power.

Pyramid power satisfies the definition of *paranormal* that I give on day one of my class: something that would violate the laws of physics if it were true, or that lacks any scientific precedent. We take a two-pronged approach, examining the claims themselves, and considering why people believe them. It's basically a course in critical thinking with paranormal examples—ghosts, bigfoot, astrology, ESP, and so on. We also bring in perception, judgment, belief formation, interpersonal and group influences, media, and culture. Along the way, we cover scientific methods including research designs, experiments, and testing hypotheses.

For their final assignment, students must write a paper evaluating a paranormal claim of their own choosing. Since his physics prof assigned something on pyramid power, Shane's newly installed skeptic's radar lights up and implants the idea for his class project: "Let's test pyramid power!"

For five-thousand-year-old dead guys, the pharaohs look pretty good. Not a day over a thousand. Pillaged royal burial chambers were discovered in the pyramids of Egypt, built for the elite of the elite. This leads some to speculate that *pyramid power* is what kept the pharaohs' remains looking so fresh.

Believers in pyramid power claim size doesn't matter. It's all about the shape. Tabletop-sized pyramids are just as capable of preserving or improving objects held inside them. It is said that stored under pyramids, food keeps longer, razor blades sharpen, and seeds germinate faster. Pyramid-shaped hats cure illnesses and sharpen the mind.[2] And, of course, in the Pharaoh-sized versions, human remains remain less gross.

Keeping your bananas and pharaohs fresh is only one of many special properties attributed to pyramids. *Pyramidologists* study their significance to architecture, engineering, culture, and archeology, not to mention those who traffic in numerology, conspiracies, biblical tie-ins, commercial products, and spiritual enlightenment.

Let's home in on one type of claim. In the 1970s, popular belief in pyramid power was just gaining traction. Writing in the *Philadelphia Inquirer*, Cindy Rose explained:

> According to a widening circle of devotees, pyramids provide new power for the people. They can sharpen razor blades, sweeten wine, cure sick plants, lessen pain, improve sex life, and boost psychic energy.[3]

And from journalist and network newscaster Edwin Newman's article in *New York Times Magazine:*

> [It is] based on the belief that the shape of the great Cheops pyramid near Cairo, when reproduced to scale, focuses energy in a psychic or mystical way that produces surprising and beneficial results. . . . The belief, among some scientists . . . is that the pyramid design creates an environment that somehow serves as a resonator altering the quality of whatever falls within its aura.[4]

Pyramid lore is less in the public eye today but still very much with us. Whenever I survey students, most are aware of some unusual beliefs associated with pyramids. No doubt the weird image on the back of every US dollar bill contributes to the aura of mystery.

YouTube is a rich source of video demonstrations and tests of pyramid power. In one video, an anonymous narrator shows off his homemade pyramid and describes his experiment.[5] His British accent makes him seem knowledgeable and legitimate. I'll call him Clive Hickenbottom-Smythe. Clive uses his scale model to store his Gillette Fusion razor between shaves. His results? He's "amazed at the smoothness and the clarity of the cut during the shave" after using the razor "many times before."

Author Praveen Mohan has millions following him on social media. His YouTube pyramid power video has over six hundred thousand views and counting.[6] His experiment uses granite containers shaped like a pyramid and a box. He looks at how well pieces of fruit and vegetables fare in the pyramid versus the box versus the open air. He also examines how well fenugreek seeds in soil germinate in the pyramid versus the box.

Each of the first five days of his experiment, Praveen opens up his containers and pulls out the contents to show the camera. On day six, he declares that the fruit and vegetable pieces from the pyramid look much better than those in the box, and both look better than the ones left in the open air.

The seeds are less cooperative. Those in the box are more sprouted. So Praveen concludes that, like Egyptian pyramids, his is better at preserving than growing.

Shane is the only student who asks to conduct an experiment for his final paper. He wants to test whether pyramid power keeps bananas fresher than a box. Although his experiment came years before Praveen's, there are some common elements. Shane sets his up at home, uses fruit, and builds his own pyramid and box.

There are also three very important differences between his method and those of Clive and Praveen.

- Shane uses whole bananas in his pyramid and box, undisturbed for five to seven days before getting judged.
- Multiple subjects provide independent judgments.
- They don't know which container held which banana.

After readying the testing room in his apartment, Shane leads one friend after another through the experiment. First, he gives brief instructions and gets their consent. Then each subject rates each banana on a ten-point freshness scale.

The experiment yields two principal results: Shane gets an "A" for the project, along with plenty of quizzical looks from his friends.

We'll get to the real findings shortly.

Backstage

The Egyptian pyramids aren't the oldest human-built things still standing, but they share a common property with all ancient structures that have stood the test of time: architectural stability. Those which endure manage to avoid the effects of erosion, gravity, vandalism, wars, and other destructive forces.

Most super-old structures are earthen mounds or rock piles, the pyramids being no exception. They've lasted because, first and foremost, they're piles of rocks that have proven hard to steal, damage, topple, or erode for five millennia. With little rain and no freeze-thaw cycles, perhaps the greatest natural danger going forward is climate change.[7]

Those on the more rational end of the pyramidology spectrum have hypothesized how ancient Egyptians could have made these structures with available technology and resources. Erich van Däniken squatted on the far opposite end of that spectrum when he wrote his mega-bestselling book *Chariots of the Gods?* Quickly discredited, he nevertheless popularized the idea that the Egyptians built their pyramids under the direction of extraterrestrial visitors.[8] But though the Giza pyramids were well-built, it doesn't mean their architects were either engineering geniuses or under the sway of alien overlords.

If you want to maximize the sheer size of a monument, a good choice is to model it off of a sand pile—perhaps an obvious choice since Giza is on the edge of Egypt's Western desert. Using large stone blocks instead of grains of sand was a good choice, too. But there's also a form of natural selection at play. The Egyptians built many kinds of structures, including several types of pyramids.[9] Only the "fittest" have survived in fairly good shape. The rest, having failed to be sandpile-like mountains of large rocks, went extinct.

The desert climate contributes to the preservation of mummies. But even more than this, the wealthy dead benefited from an array of embalming options. One replaced internal organs with sawdust and other materials, dehydrated the skin and muscles with salts, and wrapped the body in resin-soaked linens. This achieved the equivalent of "architectural stability" for human bodies. Left dry and untouched, the leathery mummy keeps its shape for thousands of years.

Over fifty reasonably well-preserved mummies have been found in Egypt alone. But only a few spent any time in the small royal chamber beneath the apex of a pyramid. The rest had more conventional final resting places, yet are well-preserved without the benefit of pyramid power. This doesn't mean that such power doesn't exist, but it does show that their preservative powers aren't special.

When the time came for Shane to apply to grad school, I suggested he consider my department. His interests fit well with our brand of research, and I wanted us to admit more students with his raw skills and enthusiasm. He applied, was accepted, and decided to come.

We started several projects as soon as he returned in the fall. One of them was an improved and expanded version of his home-brewed

THE POWER OF THE PYRAMID

pyramid experiment. To understand why we'd bother to do this, it's useful to look at some of the fatal flaws in Clive's and Praveen's experiments described earlier.

Clive built his pyramid to proper scale and aligned it with magnetic north. The razor was set on a pedestal to keep its blade at one-third the height of the pyramid.[10] But there are serious problems with the protocol he described in his video. Clive never said how many shaves he got from the blade, or how many he'd normally get without the pyramid. He also didn't say whether or not the pyramid had proven reliable with multiple razors. Finally, he was the only person rendering a judgment, not based on microscopic examination of the blade's edge but, instead, based on his subjective experience of how his shaves felt to him.

All this is to say that Clive's evidence is anecdotal—the weakest kind of evidence—and could be hopelessly biased. If pyramid power were at work, we wouldn't be able to tell from Clive's experiment.

Praveen's protocol bettered Clive's in several ways. He used multiple pieces of fruit and vegetables and some seeds in soil. He was clear about time spans, and he used both a box and an open-air condition in a way analogous to how control groups are used.

But—to borrow an ancient civilization metaphor—his experiment had an Achilles' heel. More than one, in fact.

1. Like Clive, Praveen served as the only judge and jury.

2. Praveen's unsterilized knife may have loaded bacteria and mold spores onto some of the fruit pieces.[11] Worse, each of the first several days he opened the containers for the camera, he touched every organic bit with his bare hands. He also contaminated the aging process with daily doses of fresh air. If mummies were trotted out for public viewing and pawed with unwashed hands on a regular basis, it's not likely they'd have lasted so long, either.

3. When Praveen made his final judgments in the video on day six, I truly couldn't believe my eyes. Not because the pyramid was so effective, as he claimed, but because it was so plain to see there was no obvious difference between the box and pyramid contents. If anything,

the apple slice from the pyramid looked worse than the one from the box. Yet, Praveen said the opposite. In his narration, it seemed that he may have focused on the freshest-looking *parts* of each item from the pyramid and the rottenest-looking areas of the boxed items.

4. Praveen was less biased in his judgments of the sprouted seeds. It was obvious that the pyramid seeds sprouted poorly compared to those in the box. My guess is there were slight differences in their preparation. Both sets of seeds were in soil in small glass jars. He simply may have tamped the soil more firmly in one than the other. He should repeat this experiment multiple times, randomly assigning seeds to their respective conditions.

For these reasons, Praveen's results are as invalid as Clive's. I think he made a sincere effort to test for pyramid power. But even better than sincerity would be controls for judgment bias and biological contamination.

The experiment that Shane ran for his class project corrected for these problems. He had fifteen subjects blind to the pyramid versus box conditions. He fixed the contamination problems by using whole fruit and not touching it while it ripened. His findings: no statistical difference between the judgments of bananas in the box versus the pyramid.

If Shane's experiment was so much better than others like Clive's and Praveen's, why did we go to the trouble of further enhancing it when he came to grad school? The main reason was that we wanted to have even better controls than were possible in his apartment.

We had three conditions. (1) Pyramid; (2) Box; (3) and Open-air.[12] Improvements included the following:

1. Three judgment scales instead of one, pretested for ease of use and reliability: Preserved–Unpreserved; Fresh–Rotten; Hard–Soft.

2. A panel of judges matched un-ripened bananas to be aged prior to each series of tests.

3. A larger sample was used: twenty-six subjects instead of fifteen.

4. Double-blind procedures were invoked.

Double-blinding meant that the project assistant who worked with subjects was, like those subjects, unaware of the condition associated with each banana. This prevented any information on the correct answers from being leaked to the subjects.

We did have one unexpected problem. One night, a building custodian making his rounds smelled the ripening bananas from the corridor. He used his master key to enter the lab and dispose of the offending fruit. We had to cancel some sessions, ripen another set of fruit, and post a sign asking all to respect the integrity of our laboratory materials.

Ultimately, the results confirmed that the pyramid has no special powers—at least in the eyes of human judges. In fact, the boxed bananas were rated slightly fresher than those from the pyramid. But the difference wasn't statistically significant, so we had to quell our temptation to announce we'd discovered *box power*.

As a social psychologist, pyramid power isn't something I'd normally study. It's the human factors that interest me. If I were truly interested in pyramid power, it would make a lot more sense to apply objective measures of claimed pyramid effects such as fruit oxidation, blade sharpness, or the mental acuity of those who like to wear pyramid hats. Using human judges as the only yardstick means running tedious experiments with lots of subjects and fancy blinding protocols. This is not the best way to go, but we only chose this path because we doubted others' claims about the objectivity of their judgments.

Our final experiment focused on the power of social influence rather than the power of the pyramid. We wanted to see if mere *awareness* of someone else's judgments can make people see and believe a paranormal phenomenon that simply doesn't exist.

Social impact theory says this can happen. It gives us three main factors through which we influence one another's behaviors and beliefs: the *number* of influencers, their *status* or power, and their *proximity* to the subject.[13] We tested all three.

For this study, we used the box, pyramid, and two matched bananas.[14] Subjects were informed that "some people believe pyramids have a special power to keep things fresh inside." And that's all we said about pyramid power.

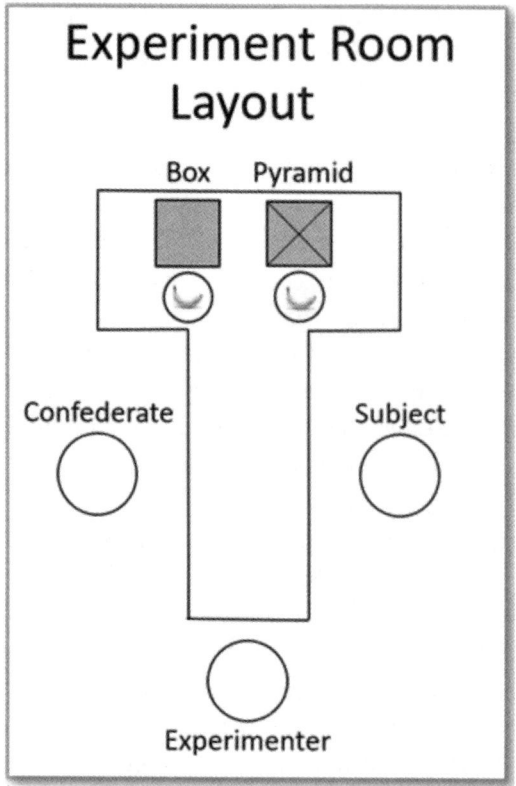

When seated in the experiment room (see the diagram), the subject was told that one of the bananas had been stored in the box sitting just behind it and the other in the pyramid. In fact, the bananas were switched after each subject, ensuring no differences on average.

The confederate across the table was a member of the project team and the source of *passive social influence*. There was never any kind of overt attempt to influence the subject. The confederate and subject would alternate voicing their freshness judgments to the experimenter for each piece of fruit. The confederate gave the same responses in every session, *all indicating that the pyramid banana looked fresher.*

THE POWER OF THE PYRAMID

We ran four conditions:

1. Subject was alone;
2. Confederate not present, but experimenter reports their ratings;
3. Confederate present at the table; and
4. High-status confederate at the table.

In Condition 2, the experimenter read the ratings supposedly given by a previous subject.

In Condition 3, the confederate was present and gave the same ratings as the "absent confederate" in Condition 2.

In Condition 4, I acted as confederate, giving the same ratings as in Conditions 2 and 3. I had higher status by virtue of being a professor while the real subject was always a student. I explained, "They recruited subjects from my class, and I wanted to see the study for myself."

These four conditions allowed us to test all three social impact theory factors.

Here are the results, by predicted effect:

- *Number:* Compared to working alone, the one confederate was enough to have a significant impact on the subjects' judgments. Most social influence studies have multiple influencers, so it's important to know that as few as one can have such an effect.
- *Status:* The high-status confederate had more influence on subjects' freshness ratings than *any* other condition.
- *Proximity:* Contrary to the theory, merely knowing about someone else's judgments had just as much impact as when the confederate was sitting there across the table.

This last result is fascinating. Merely *hearing secondhand* that a random, unknown person saw something paranormal influenced subjects as much as having a real person right there in front of them say the same thing. It may help to explain why, according to national surveys, so many people have so many paranormal beliefs.[15]

After the judgment procedure, the subject was left alone to complete a short questionnaire. This measured perceived differences between the two bananas, only this time without any possible social pressure. The questionnaire showed that subjects didn't simply go along with confederates. They actually *believed* the pyramid banana looked better than the boxed banana in the confederate conditions. The socially implanted false belief had become internalized.

Our experiment proved to be more than just a fun demonstration project. It's been useful as a teaching tool—especially in the Sociology of the Paranormal course, a version of which Shane eventually developed and teaches to this day. We also published two articles from the project: one technical, and one for a broader readership.[16]

Shane has gone on to become a very successful sociologist with a string of books and publications in top journals. I'm proud to be his friend, to have been on the same faculty with him for nineteen years, and to be able to say "I knew him when. . . ." It's true. I knew him when he first combined fruit and cardboard with critical thinking and made some original science.

CHAPTER FOURTEEN

ESP: Can It Be?

If it's the psychic network, why do they need a phone number?
—Robin Williams[1]

Frontstage

One wintry Midwestern weekend, I set out early in the morning to spend a day at the big fair two hours away. Not the famous Iowa State Fair, with its life-sized butter cow and batter-fried Oreo cookies on a stick. That one's at the opposite time of year in the humidity-soaked days of summer. No, I'm going to the big psychic fair in Des Moines, Iowa. Its full title is *The All-Iowa Holistic and Psychic Extravaganza!*

I pay a small admission fee and enter a cavernous and chaotic exhibition hall. From a slightly elevated vantage point near the entrance, I can see dozens, if not hundreds, of neatly arrayed booths and displays. I set off on a random walk, like a kid in a paranormal candy store. My *journey of personal growth and enlightenment*, as promised in the fair's newspaper ad, has begun.

Within a few minutes, I notice everything I encounter seems to fall into a small number of categories:

Paranormal services: Tarot and psychic readings, pet psychics, astrologers, Reiki practitioners, crystal healers, quantum touch therapists, aura readers, and more.

Vanity items: Specially formulated creams and lotions; makeup, hair, and nail products. Most are "energetically activated" or "vibrationally tuned."

Energy products: Crystals in the raw or fashioned into jewelry and charms. Therapeutic magnets, aromatherapy oils, sound therapy recordings, and Feng-shui items.

The Arts: Paintings with spiritual and paranormal themes, drawings, sculptures, jewelry, poetry books, audio recordings, and handcrafted instruments.

I buy a pack of colorful brushed aluminum coasters because they remind me of the shiny beverage tumblers we had when I was a kid. Gas stations gave them away with fill-ups. Aside from the nostalgia value, the one-dollar package claimed that when you set a drink on one of their coasters, it will "align energy frequencies to center the spirit, support good health, and promote happiness." Heck, I can't afford *not* to buy them.

What I really want to take away from the fair is a personal reading from one of the many psychics. I scope out a dozen or so booths, all charging around five dollars. Some of the psychics spend longer with clients than others. Some also look more professional, realizing it doesn't help business to have a cigarette dangling from their lips, or a filthy, heavily tasseled, faux-exotic tablecloth and red-lit lamp.

"Dave the Psychic" catches my mind's eye. I stop to chat with some of his fans. They all vouch, "He's the real deal!" Some are return customers. I sign up on the clipboard for the next available slot. Fourth in line.

I watch as Dave sits with each client for ten minutes or so. Judging by the way he sometimes leans in to the client and lowers his voice to a whisper, it appears he gets very personal with them. My anticipation builds over the next thirty minutes.

My turn finally arrives, and I take my seat across from Dave. Though I'm a little shy, he quickly puts me at ease.

"Hello there! What's your name?" he asks, smiling. I so desperately wanted to reply, "You're the psychic. You tell me!" Not wanting to get off on the wrong foot, I choose a less snarky path.

"Barry. What's yours?" I ask back, making it obvious I'm staring over his shoulder at his huge *Dave the Psychic* sign. He gets it, chuckles, then begins with a soothing voice.

"So Barry, this is simple. No gimmicks. No tea leaves or crystal balls. I use my five senses and my second sight...." He makes air quotes, adding "... whatever that is." I appreciate the self-deprecation. He continues.

"With these tools, we'll exchange energy and get into sync. You'll feel yourself relax a little, but don't worry. It's not hypnosis or anything like that. Your heart and mind will let down their guards and tell me a story. I'll read it back to you as it unfolds before me. Some of it may resonate perfectly. Other parts might seem confusing or surprising. The more you remain open, the better the chances we'll reveal some truths. Sound okay?"

"Absolutely!" I reply.

"Alright, Barry. Let's begin."

Trying to relax, I pull my padded folding chair closer to the table and lean back into it.

Dave readies himself with some slow, deep breaths. His eyes do this weird thing, rolling up in his skull while his lids flutter for a few seconds. He slumps back in his chair, his hands in his lap, and his gaze now fixed on me.

A long, silent minute passes.

"Are you recovering from an injury?" he asks.

"No, I don't think so."

"But I sense something in your leg . . . your knee. I see you limping. Not today, but I sense a limp. A scar? Past injury?"

"Oh yes, I tore my knees up good last fall on a gravelly softball field. They're always bandaged and scabby. Still not healed. You can see that?"

"I see *something*. But now I'm getting something else, something more important. You're worried," he declares. "Is it your family?"

"Uhm, yes, probably."

"Hmmmm. Your marriage . . ." he's scanning me intently. I can feel my eyes open wider in anticipation. ". . . feels positive. But there's a child. You've got one . . . that's . . . teenaged? Around?"

"That's right. Twelve."

"Yeah, at that age, the boys are still all goofy and innocent, but the girls are starting to try to grow up too fast. Right?"

"Oh, man. So true."

"You've got a girl."

"Uh-huh."

"Older boys are showing interest. You want to protect her, but you don't want to smother her . . . to dim her light. . . ." He trails off. "But there's also something happening at work that worries you. You're a professional. Special training. A lot invested. Now you see some people leapfrogging over you. Maybe not fully deserving. Does this make sense?"

It brings to mind all my rejected article submissions to journals while, at the same time, they publish all sorts of shoddy work. Publications advance careers, so yeah, in a way, it does sometimes feel like being left behind. I'm only human.

"Definitely," I tell Dave.

"Don't let that get to you. Sometimes things seem unfair on the surface, but there's a bigger picture. Those people pay the price. Some of them hate their jobs or alienate loved ones. Some waste valuable time in turf battles. Does this make sense?"

"It does." I sense the clock running, so I ask, "Can we switch topics?"

"Of course, Barry. Anything."

"I miss my mom and dad. How do I deal with that? Will it get any easier?"

"Losing a parent is incredibly tough," he says comfortingly. "Losing both can be devastating. You feel disconnected sometimes. Incomplete. But it makes you appreciate your connections with your partner and your daughter. Any other kids?"

"No."

"Cherish your little girl. Cherish both the females in your life and all the truly valuable things. The memories of your parents—I see them in a good place. I sense them through you. I sense their pride and love, although they may not have told you enough."

He's right that my parents weren't expressive toward me. Sometimes one of their friends or a relative would tell me how Mom or Dad went on and on about how proud they were of me and the way I was turning out. That was enough for me. For better or for worse, their parents raised them that way. And I may have overcompensated the other way with my daughter, showering the poor girl with love at every opportunity, whether or not she wants it.

And that's why, with the clock running out and Dave making me think about parents, love, and children, I get misty. A tear forms at the corner of my eye and rolls down my cheek. From all appearances, I look like any other emotionally satisfied customer.

Except I'm not.

Dave confirmed all my suspicions: He's an obvious fraud.

Backstage

Claims about *extra-sensory perception,* or ESP, evoke a spectrum of opinions ranging from certainty that it exists to certainty that it doesn't. Opinion polls over the last two decades show that anywhere from 40 to 60 percent of Americans believe in ESP, plus many more who say they're unsure or open to the possibility.[2]

Most people base their opinions about ESP on firsthand experiences, others on testimonials and its popularity in mass media. However, these aren't reliable sources, and most people are unaware of the large body of published research on ESP.

Those who are most knowledgeable about the research tend to fall into one of three camps: (1) a relatively small group who devote their careers to the study of ESP and believe firmly the evidence supports it,[3] (2) those who are highly doubtful because, after a century and a half of ESP research, they find the evidence unconvincing, and (3) a subset of ESP agnostics who, for a variety of reasons, sit on the fence and hold firmly to that position.[4]

The 1984 blockbuster movie *Ghostbusters* didn't do ESP research any favors. In a scene with Bill Murray as Dr. Peter Venkman, we see him conduct an ESP test at Columbia University's (fictitious) Paranormal Studies Laboratory.[5] Two caricatured students are seated across the table from him: a gullible, attractive blond female and a frizzy-haired male nerd. They're subjects in a psychic guessing game. In it, Venkman holds up cards with symbols hidden from his subjects but visible to us looking over his shoulder.

"All right," he instructs the male, "I want you to concentrate and tell me what card you think it is." We can see it's a star. The subject pauses in concentration.

"Square!" he responds confidently.

"Good guess, but wrong." Venkman shows him the star, then flips a switch on a small console. We hear the loud *BZZZZZZT* as it delivers a shock, making the young man jerk and grunt.

Venkman looks up at the female subject. With an inappropriately flirtatious grin, he holds up the next card: a circle.

After deep contemplation, she asks, "Is it a star?"

"It *is* a star!" extols the unethical parapsychologist. She's delighted. He returns the card to his table without showing her the other side.

The deceptions continue in this way for a couple more rounds. The male gets more agitated with every shock until, finally, he angrily stands up and quits the study.

"But you volunteered!" Venkman protests—as if that granted permission for the torture. The subject storms out the door.

Venkman goes around the table, sits very close to the female subject, and places a consoling hand on her arm.

"You'd better get used to that," he says with fake empathy. "It's the kind of resentment that your ability is going to provoke in some people."

If parapsychologists find the evidence for ESP supportive and compelling but scientific skeptics look at the same evidence and find it lacking, what should we believe?

Let's start with the assumption that anyone—believer, skeptic, you, or I—can become deeply convinced of something and be flat-out wrong about it. Look no further than politics, religion, or conspiracy theories. Within these realms, we find people who fervently believe that particular claims are true, while others just as strongly believe the opposite. One side certainly has it wrong. Possibly both.[6]

So it is with ESP. Parapsychologists have told me it's incomprehensible to them how anyone could look at the evidence and still be skeptical. But the staunchest skeptics feel just as sure that believers are only fooling themselves.

ESP: CAN IT BE?

Shouldn't it be true that, if anyone can tell us if ESP is real, it's the parapsychologists? If they're devoting careers to testing ESP claims, drawing from the same experimental and statistical methods as mainstream science, why don't we believe them?[7]

Let's take a closer look. First off, some definitions. The most commonly studied forms of ESP are:

- *Clairvoyance*: acquiring information by means other than the known senses
- *Telepathy*: communicating thoughts by means other than the known senses
- *Precognition*: predicting future events beyond normal abilities to do so

Often a fourth category is lumped in:

- *Psychokinesis*: intentionally affecting matter using only the mind

To study any form of ESP scientifically, its existence must be well-established, and it should be distinguishable from other phenomena. This is where the problems begin. Parapsychologists take for granted that these criteria are met. Skeptics don't, for several reasons. First, there have been many ESP experiments that appeared to demonstrate a psychic phenomenon. But when looked at more closely, they were shown to be tainted by bias, methodological flaws, fraud, or a failure to replicate in retests.

Second, even when researchers think they've experimentally isolated one kind of ESP, they generally can't be sure *which* kind. Imagine a card-guessing experiment like in *Ghostbusters*—without the shocks, cheating, and sexism. If subjects reliably scored above the 20 percent chance rate—25 percent, let's say—that would support a *clairvoyance* hypothesis. Subjects seemed to access information not available to the normal senses. But if the experimenter or anyone else knew what cards the subjects were seeing, subjects could have read their mind using *telepathy*. Or subjects could have used *precognition* to see their future results and adjust their responses accordingly. As long as we're speculating, subjects could have used *psychokinesis* to bias the card shuffle in their favor.

The experiment portrayed in *Ghostbusters* wasn't totally fictitious. It was modeled on the work of Joseph Banks Rhine in the early 1930s. A botanist by training, Rhine wanted to establish parapsychology as a scientific subfield of psychology. He introduced *Zener cards* into ESP test protocols, just like those used by Venkman in the film. Named for their inventor, Karl Zener, the standard deck included twenty-five cards, five of each symbol: star, square, wavy lines, cross, and circle. In the simplest study, the experimenter shuffled the deck, and subjects tried to guess which symbol appeared on a card in each trial. Some subjects seemed to be able to guess the symbols at above-chance levels.

As a plant guy, Rhine might be forgiven for not anticipating the challenges of studying people. Unlike petunias, human subjects often want to please the experimenter and respond to subtle cues that may help them do so. For instance, in some experiments, sharp-eyed subjects could detect faint impressions of the Zener symbols on the backs of the cards. In other studies, the experimenter's glasses reflected the symbols off the cards as he read them to the subject. There's also the opportunity for experimenters to cue subjects with eye movements or facial expressions.[8]

These were relatively simple and solvable problems: Ensure the symbols aren't visible through the cards. Eliminate reflections. Use a machine to shuffle the deck after each trial to eliminate card-counting and biased shuffling. Hide the experimenter behind a screen. As researchers implemented these fixes, success rates eventually fell to chance levels and researchers mostly abandoned Zener cards entirely—though everyone else can still buy them on Amazon and conduct their own poorly designed ESP tests at home.

Rhine's work set off a repeated cycle spanning most of the last century.[9] Every five to ten years, someone publishes a creative new study that seems to demonstrate once and for all the existence of some form of ESP. The spooky findings create a big stir in the field and get widely disseminated in mass media. Some examples:

Psychic Spies! In a CIA-funded project, subjects sketched scenes viewed by "senders" at faraway locations. One subject seemed to produce uncannily accurate sketches and received much publicity. But problems were later identified by skeptics: Only a single judge evaluated the sketches, and the criteria for success were loose and subjective. The CIA concluded that the study produced no useful data.[10]

Reincarnated! The most famous claims promoting reincarnation involve children who offered verifiable details from the lives of dead people they never knew.[11] Skeptics have investigated all of the "best cases." Invariably, they find serious shortcomings in the methods used by investigators.[12]

I Know You're Staring at Me! Biochemist and parapsychology researcher Rupert Sheldrake (and others) conducted studies in which subjects allegedly reported when someone outside their field of view was looking at them. The effect was small but statistically significant. There are two big problems, however. First, the methods used were open to various biases. Second, the results didn't replicate when others conducted the experiments.[13]

The Future Is Now! Subjects read a list of words. When asked to recall as many as possible, they more accurately remembered the words they would be seeing later. Pause for a second and try to wrap your mind around that one. Here's a little more about it.

Psychologist Daryl Bem of Cornell University published this study (with several others) in a major psychology journal. It had three parts: (1) a computer screen presented a series of words to each subject; (2) they tried to recall as many of those words as possible; and (3) the computer randomly chose half of the original words and showed them to the subjects, who then "practiced" by reading, sorting, and typing them into the computer.

When Bem tallied the test results from the word-recall task, he found that subjects had slightly better recall for the practice words, even though they hadn't yet practiced them.[14]

If this were true, it would be earthshakingly amazing. Bem even cooperated with others in attempts to repeat the findings. But the results didn't replicate. No time-bending effects were found.[15] While the original experiment appeared in popular news media worldwide, the failed replication attempts did not. Skeptics could have predicted this outcome, even without the benefit of ESP.

As with any ESP research, the results would have been stunning if they'd held up to replication. For starters, physics, physiology, and psychology would need major overhauls to accommodate ESP's implications: that minds exist without brains; thoughts directly influence objects; mind-to-mind communication is unbounded by distance; time can run backward. It would be an exciting new era for science but not one we'd want to rush into based only on a few intriguing studies.

It's fun to think about the possibility of information flowing back from the future. Wouldn't it be great to anticipate a friend's reaction to a birthday gift? Or your future child's forthcoming outcomes in life? You could even avoid calamities. Suppose you envision your car getting T-boned next Thursday at the corner of Pine Street and Cypress Avenue. Naturally, you would keep away from that spot when the fateful day arrives. But if the accident doesn't happen, how could information about it have gone back in time and allowed you to foresee it?

Some have speculated that both futures happen—one with the accident and one without—and send information back in time so we may choose which we'd prefer to enact. But if two futures can play themselves out for us, why not three, a hundred, or an infinite number of them containing every conceivable scenario with infinite variations? It gets messy. You'd have to choose your route knowing that even if you avoid the T-bone accident, other futures will hit you at the next intersection with a meteor, tornado, gunfire, none of the above, or everything all at once.

While some kinds of ESP imply time nonlinearity, others imply that souls can survive death and that disembodied minds are all around us, flitting about like immortal butterflies. What if our essences are unknowingly intertwined through a vast collective consciousness? What if quantum-level connections between matter and energy mean we are literally one with everything?

ESP: CAN IT BE?

Why does science slam its door in the face of such beautiful possibilities?

It doesn't. While individual scientists may be overly quick to dismiss new ideas, open-mindedness is baked into the enterprise of science. But not every idea that comes down the pike is automatically credible. Science requires *really good reasons* to believe claims. Good reasons come by way of *theories* that provide clear, logical explanations, and *evidence* that provides unambiguous observations to support theories.

Science has always been open to radical new ideas. We've seen this with crazy-sounding claims that eventually were accepted by science. Examples include global geography shaped by continental drift; peptic ulcers caused by bacteria rather than stress or spicy food; and Einstein's space-warping general relativity theory superseding Isaac Newton's gravitational theory. But it can take decades to gather a body of evidence sufficient to topple prevailing knowledge. Appropriately, it's not enough to offer a weak body of evidence and mere speculation in lieu of a rigorous explanation. The usual standards would have to be relaxed *a lot* for science to deem that ESP is adequately demonstrated and explained.

Here's the crux: ESP claims could be true, and science's current theories may be wrong. But to date, for *every single one* of parapsychology's extraordinary claims, careful detective work eventually revealed potential flaws in the research or statistics. As in Bem's studies, if you fix the problems, the evidence for ESP vanishes. But for reasons more sociological than scientific, parapsychology soldiers on. Arthur Reber and James Alcock described its persistence in the face of endless disappointments:

> As excitement about each new procedure wanes, a resurgence of interest develops when another, apparently successful procedure is reported. . . . The single feature that marks this extended period of research involving literally thousands of published papers, hundreds of conferences and meetings, and dozens of review volumes is that *nothing has been learned*.[16]

If you're a parapsychologist trying to convince the unconvinced, it's an uphill battle—as it should be. You've no theory to explain ESP, no method to produce it reliably, and you can't necessarily distinguish one kind from another.

That parapsychologists endure in the face of harsh criticism makes them neither foolish nor virtuous. They're humans. They have beliefs, skills, egos, and passions. I've met and corresponded with dozens of parapsychologists. Most are sophisticated and sincere researchers. Some are probably biased by overenthusiasm, but this is true in any field. Skeptics can't prove the *non*existence of ESP (or of paisley-colored elephants), so it's always possible to hope that the dogged pursuit of reliable evidence will eventually prevail. But for skeptics, the lush oasis of proof that parapsychologists see looming on the research horizon is only a mirage.

Who are these skeptics?

Among the strongest proponents of parapsychology are parapsychologists, by which I mean scholars for whom ESP is their main research focus. In contrast, virtually all of the most prominent skeptics of parapsychology have other interests. Some are magicians or psychologists with expertise in illusion and deception. Others include biologists, geologists, astronomers, statisticians, private investigators, photographic and videography experts, and philosophers of science.

There are a few skeptics who've devoted much of their careers to evaluating ESP research. Susan Blackmore is one of the most interesting of this lot. A dyed-in-the-wool parapsychologist early in her career, she tells the story of her conversion in her autobiography *In Search of the Light*.[17] Hard as she tried, she couldn't replicate her mentors' evidence for ESP. At first, she thought there must be something wrong with her methods. But the more she tightened her research procedures, the clearer it became that ESP wasn't happening and that her mentors' studies were the ones with the holes in them.

Ray Hyman is one of the most skilled and prolific ESP skeptics. He also started out as a believer and actually earned some money working as a palm reader, routinely getting wildly positive feedback from clients. He tells the story of how one day, on a whim, he decided to see what would happen if he purposely gave false readings. To his shock, these clients were no less impressed. The experience contributed to his decision to major in psychology.[18]

Richard Wiseman is a British psychologist and author of many books and articles on ESP. He's among the most active skeptics insofar as public engagement through his media presence, YouTube videos, speaking appearances, and articles in mass media outlets. This is all in addition to conducting and publishing research on several fronts.

It's true that ESP skeptics tend not to be trained by parapsychologists. But their specialized knowledge of experimental methods, misperception, and statistics makes them ideally suited to evaluate ESP research. Unfortunately, at times, some skeptics can be dismissive and mean-spirited—as are some parapsychologists toward skeptics. But mostly, the skeptics speak up because they feel a professional responsibility to look into parapsychology's extraordinary claims.

Despite all the back-and-forth, *very few skeptics or proponents ever change their minds about ESP.* This is fascinating to me. Both sides know how to test claims and evaluate evidence. Both sides also know about judgment biases stemming from overattachment to a belief. Despite this, knowing that both groups can't be right, each concludes the other is self-deluded.

I conducted a survey to help understand why parapsychologists and skeptics were so far apart in their beliefs about the evidence for ESP.[19] I decided that I should ask them directly. I got the ball rolling by contacting a couple dozen parapsychologists and skeptics whose work I knew. I asked them who, excluding themselves, I should include in my sample of respondents. Then I contacted those suggested people with the same question. After several rounds, responses converged on a list of the most renowned and accomplished skeptics and parapsychologists.

Next, my research assistants and I visited with parapsychologists and skeptics at conferences, laboratories, universities, and homes. We conducted forty-six interviews, including thirty-one well-known parapsychologists and fifteen top skeptics.

We covered a wide range of topics, but here I'll only talk about a few main differences that emerged between the most elite of the proponents and skeptics.

The two groups were similar in most ways, such as their scientific interests and credentials. However, differences emerged when we talked about their earliest interests in ESP: Only the proponents in their youth were drawn to the writings of parapsychologists. None of the skeptics were. However, half the skeptics became interested in ESP via magic and magicians. Prestidigitation planted the notion that misperceptions can lead to paranormal beliefs.

This was certainly true for me. A curious kid, I saved up my allowance and bought a copy of the encyclopedic *Scarne's Tricks*.[20] Today, it remains in arm's reach at my desk. I haven't learned too many tricks, but it showed me many of the simple deceits behind seemingly impossible feats of magic.

Always looking for a social angle, I asked respondents who they talked with about their ESP interests. Proponents and skeptics spoke equally to students and others in their professional networks. But proponents were more than twice as likely as skeptics to discuss ESP with personal contacts such as family and friends. The proponents appeared to be more personally invested in their ESP work than the skeptics. Having a personal stake in one's research is a concern in science. It creates the potential for bias, in this case, one that existed for proponents but not for skeptics.

Proponents and skeptics had quite different opinions about parapsychology's biggest problem. Proponents felt they needed more research and improved methods. In contrast, skeptics wanted to see a theory—a rigorous account of *how* ESP works if it truly exists. This helps explain why, according to proponents, skeptics are never satisfied with research findings.

The lack of theory means parapsychology lacks a scientific foundation, despite sharing some research methods with legitimate sciences. There always will be new research findings to tantalize proponents and skeptics alike. But without that theory, few skeptics will be swayed, and ESP research will remain on the fringe.

Finally, I asked respondents to compare their current acceptance of ESP to how they felt about it at the time of their earliest professional interest. Skeptics and proponents diverged early in their careers. Not a surprise. But what happened as they became more familiar with the fuller body of ESP research? Did such knowledge soften the skeptics' resistance to the possibility of ESP? Were proponents' beliefs weakened upon learning about decades of marginal results and the complete absence of theoretical progress?

Neither group softened its position at all. On the contrary, both became more extreme: Proponents grew surer that ESP is real, and skeptics grew more skeptical.

There are reasons beyond psychological for the persistence and hardening of beliefs in the extraordinary. All the professionals we interviewed were embedded in networks of supportive people who shared their beliefs. We've all felt that rush of self-validation when interacting with like-minded people. It could be chatting at a quiet gathering, swaying with the crowd at a concert, or cheering at a sporting event. This helps to account for why parapsychologists take for granted that ESP exists. To skeptical me, the wholehearted acceptance of ESP seems unsupportable. It stretches the belief in ESP far beyond what the evidence actually supports. But would skeptics reverse their position if a well-formulated theory and reliable evidence issues from parapsychology tomorrow morning? I'm skeptical of this, too. Most scientific revolutions must wait for a new generation of scholars to rise through the ranks and lead their field past outmoded beliefs.[21]

If psychic fair psychics like Dave had the gift of precognition, parapsychologists would be on them like honey on a hot biscuit. Psychics would finally have captured the illusive validation for ESP sought by parapsychologists. In return, parapsychologists would provide the psychics with an invaluable scientific endorsement. But the parapsychologists know better. Dave the Psychic wasn't making evidence. He was making a living. My brief session with him revealed several common tricks of the trade.[22]

Dave gave me what's called a *cold reading*. It's when the psychic has no prior information about a first-time customer.[23] Sessions typically start with the psychic trying to bond with the client. Helping them feel relaxed and affirmed establishes rapport and trust. Dave was really good at this. He was affable, made eye contact, smiled, asked questions, cared about the answers, called me by name, leaned in, and spoke softly like he was confiding in an old friend. He made me and every client feel special.

Psychics know they'll get more wiggle room when the client likes them. To like is to trust.[24] So when he asks you to keep an open mind if something doesn't make sense at first, you trust that it *does* make sense, and it's up to you to figure it out. Advantage Dave.

The fallibility of human memory is another factor that works to the psychic's advantage. Most people's memories of their readings are selective. They immediately move past the false or ambiguous pronouncements and key in on correct information the psychic would appear to have had no way of knowing. The natural bias toward looking only for confirming information means the psychic can never appear to be wrong.

Psychics also know subtle ways to extract information from clients. One effective technique is to repeat things the client said earlier, disguising them as new revelations. The client might mention a brother who died in an accident. Ten minutes later, with the client's mind reeling from being jerked in all sorts of directions, the psychic declares:

> Someone very close to you, a male, passed away tragically—here one minute, gone the next. Do you understand what I'm saying? Does this make sense to you?

The client immediately thinks of her brother, gets teary-eyed, and says, "Yes."

Score another credibility boost for the psychic.

There's a hidden purpose to the psychic's question, "Does this make sense to you?," which capitalizes on its double-meaning. In fact, he can tag the question onto *any* psychic claim, and nearly all clients will answer "Yes" every time. Why wouldn't they? The *question* makes sense whether or not there's a dead brother. Even if there isn't, the client's positive answer to the question makes it look like the psychic knew what he was talking about.

You may already have seen through another part of the psychic's verbal trickery: "Here one minute, gone the next." Many people die suddenly from heart attacks, strokes, and accidents. Were that the case for a client's loved one, the fake precision makes the psychic appear all the more convincing. But even if the deceased suffered a long, lingering decline, we usually think of the moment of death as a transition that happens at a specific time: the moment when respiration and cardiac activity stop for good.[25] Before that time, the person was here. After, gone. The psychic's vague claim still looks like a psychic "hit."

Another trick is widening the claim's timeline to make it look correct. When Dave asked if I had an injury, a "yes" from me would have counted as a hit. Instead, my "no" made him branch off into the past—and who *hasn't* been injured at some time in the past? But even if I'd never been injured, there's always the future. The simplest way to turn a miss into a hit is to move the event forward in time and confidently warn the client to be careful of the injury looming in their future.

Any psychic worth their salt can interpret a variety of nonverbal cues. Clients convey volumes of information through their posture, gestures, clothing, age, physical condition, jewelry, vocal intonations, vocabulary, and facial expressions. Any combination of these—especially my age at the time—could have served as the basis for Dave's forays into my personal and work lives.

Dave confidently declared, "You've got a girl," without my having revealed my child's gender. Most clients would recall that statement and say, "He *knew* I had a daughter!" In fact, he'd *asked* if I had a child, and I said I did. He *asked* about my child's age, and I told him that, too. Then he described some differences between boys and girls of that age, and *asked* me to respond when he said, "Right?" The trick was that his gendered descriptions would typically evoke different reactions from a parent: bemusement toward a goofy son versus the deeply furrowed concerns known all too well by fathers of daughters on the cusp of puberty trying to look and act older than they are.

One surefire way to pull back the curtain on psychics' trickery is to resist the urge to correct them if they go astray. This happened when we talked about my parents.

I was being honest when I said I missed them, but it wasn't because they were dead. It was because they lived twelve hundred miles away, and I hadn't seen them in a year. Dave inferred their premature demise based on my age, my mention of sadness, and the chance that I might have raised the topic because I wanted to know if their souls were in a better place than the Boston suburbs. Telling me "They love you" is standard fare for psychics who claim to channel information from dead family members. They know that telling clients what they long to hear triggers emotional responses, softens critical thinking, and fosters positive feelings toward the psychic. All great for repeat business.

Despite seeing through Dave's rhetorical trickery, he was still able to play me like a cheap fiddle. He successfully triggered an emotional response by getting me thinking about my relationship with my parents. Still, the experience was enlightening. I can't even imagine how potent his effect might have been were my parents deceased at the time, and had I believed in his ability to connect with them.

Is what Dave does a good thing or a bad thing?

On the one hand, psychics tap into feelings of connection and powers beyond our normal capacities. Clearly, this is something many people desire. Plus, psychics don't usually prescribe potentially toxic remedies, and most charge modestly for their services.[26] Clients lay cash on the table, converse with a careful listener, receive thoughtful advice, and leave their sessions feeling better than when they arrived.

Seems like a good thing.

On the other hand, it's all but certain that psychics aren't psychic. It's far more likely they're giving advice akin to a regular therapist's—although without proper training, ethical boundaries, or any real understanding of their client. All their feats are doable with age-old tricks. By this view, psychics commit malpractice and consumer fraud with every session.

Seems like a bad thing.

Faking communication from the dead is a huge ethical no-no for me. It preys on emotionally vulnerable people and potentially does real psychological harm despite the feel-good glow clients receive in the moment. There are proven methods to help the bereaved manage grief and move forward with some degree of closure. But if you're under the sway of a psychic claiming to connect you with your people in the Great Beyond, this is not dealing with grief effectively. It's keeping the client in denial—the opposite of closure. It's also lying to people to get their money.

Seems like a *very* bad thing.

ESP: CAN IT BE?

My advice: *Consumer, be informed; psychic, be honest.* Idealistic, I know. But ask yourself two things: First, would you go to a psychic if you knew the tricks they use and the scams they perpetrate?[27] Second, would you go if your psychic informed you up front that he's not *really* psychic? If you proceed, either option negates the deception and fraud. And it wouldn't necessarily put them out of business, either. Mentalists are nothing more than magicians who do cool psychic tricks. They're entertainers and don't claim to have psychic powers—although the best of them truly seem to have mastered ESP. But they won't pretend to be your life coach or call forth your granddad from the great beyond. They're just fun.

How I wish the fake psychic's services were real. I'd happily wait all day and night in line. I'd pay virtually any fee to have a catch-up call with my now deceased parents.

How could you ever put a price tag on such a thing?

To answer that question, just ask Dave.

CHAPTER FIFTEEN

Are We *Real* Doctors?

Some of them wanted to sell me snake oil. I'm not necessarily going to dismiss all of them as I have never found a rusty snake.
—Terry Pratchett[1]

FRONTSTAGE

Dr. Marcia Rowan is a great date, but I think she may be killing people.

I've been dipping my toe in the online dating pool for the last four months since the end of a long relationship. My virtual Cupid's algorithms offer Marcia as a match, and her profile really does stand out. Lots of plusses, no red flags. I reach out, she responds. A little more messaging, and a few days later, we're meeting at a sidewalk café.

All goes well over coffee. She's quick-witted and intelligent. Confident but soft-spoken. A very, very nice person. I like that she's a doctor and that she's tall, athletic, and attractive.

"Do you have any kids?" Asking this question is in the top five Unwritten Rules of Dating.

"I have two," she says, showing her pride with a shift in posture. "One's a geologist, the other's in a nursing program. Both are out making it on their own. You?"

"One daughter. She's bopping around California trying to figure out what to do with her life. How are you liking your empty nest?"

"Not so bad! I heard a joke I can relate to: 'Having children is like living in a frat house. Always noisy, everything gets broken, and there's a lot of vomiting.' So no, I don't miss the old nest."

I'm on the verge of snorting with laughter.

We make plans to meet again.

A few days later, we're sipping reds at a wine bar. Marcia asks about my job, and I give the thirty-second elevator summary. I ask her about the medical practice she mentioned in her dating profile and alluded to briefly at our short coffee date.

"I'm a naturopathic physician." She discloses this while locked onto my eyes to gauge my reaction.[2]

They say that in a car accident, everything seems to happen in slow motion. The same can be true when dating goes wrong—both the slow motion and the crash at the end. When Marcia says what she does, it's like the road we're driving on starts to slicken with winter rain. No immediate danger, but it is enough to activate some caution. My thoughts are racing while trying to hold eye contact, trying not to look distant while nodding and smiling, my brain holding a conversation with itself.

She's not a real doctor. . . . Neither are you, sociologist!
And I like her. . . . Then hide your disappointment and get over it.
Maybe naturopathy really works. . . . But you think it's pseudoscience.
She's really pretty! . . . Then try to keep your mind open.
I think I really like her! . . . Better read up on naturopathy.

"Mind telling me more about it?" is what comes out.

"Sure. Four years post-graduate training in different treatment modalities. Practicing for over twenty years. Leadership role in the Women's Health Network. I publish, co-edit a journal . . . uhm . . . what else can I tell you?"

I ask, "How did you get interested in it?"

"I've always loved science," she says.

Good answer.

My problem is that I've read articles where naturopathy gets criticized along with unproven and dangerous treatments. I'm not sure what to believe. All I know for sure is that the lovely Dr. Rowan is sitting right here, right now. To whatever degree I'm biased against her life's work, her gentle presence keeps my mind open. It's hard to believe she's a danger to anyone.

She continues, "But I have some major beefs with conventional medicine."

"For instance?"

"Regular physicians overmedicate and don't get to know their patients. They bring out the big pharmacological guns to treat symptoms but ignore the whole person. It's inhumane."

Maybe she's overgeneralizing, but who hasn't encountered *that* doctor? Usually a male, he makes you wait an hour, acts like you're wasting his time for the minute he grants you, enters a prescription on his laptop, and walks away.

"What does it mean to treat 'the whole person'?" I ask.

"We're trained to listen. I get to know patients. We talk about what's going on in their relationships, their diets, their minds. I'll only prescribe treatments that support natural healing. We work with the body's own systems for repairing itself, eliminating toxins, striving for balance."

"What kinds of treatments are you talking about?" I ask.

"We have a lot of discretion. We can choose from a whole array. Diet, herbs and supplements, vitamins, homeopathy, detoxes, acupuncture, and others you probably haven't heard of."

Skeptical pressure is building inside. I release a little by hissing out, "Is it science?"

"Mostly," she replies, "and there's a lot of clinical research to back it up."

"Do you think *all* of these kinds of treatments are effective?"

"Well, no. Not everything works for everyone, but that's why we get to know our patients as whole people. We tailor the treatment for each person's systems."

I try to clarify what I'm getting at: "I'm just wondering how much these treatments might work like placebos." She could take this as me implying there's literally nothing to naturopathy. Instead, she takes it well.

"Sure, there's an element of placebo that aids healing, just like in conventional medicine. But I gravitate toward treatments with solid backing from clinical research."

"Such as?" I ask.

"Acupuncture, herbs, homeopathy, diet." She describes amazing healings of obstinate illnesses. Borderline miracles by the sounds of them.

"That's great, but obviously it's 'anecdotal.'" She has to know that such evidence is notoriously unreliable in medicine and should be taken with a grain of salt. Maybe these patients were being treated at the same time by conventional doctors. Who knows?

"You said there's scientific evidence for the things you prescribe. Can you point me to anything you think is especially convincing? For *any* of the treatments you use?"

"Sure!" She sounds pleased I'm interested enough to ask. "I'll send you some."

Switching gears, we explore a few other topics and get to know each other a little better. She talks about her family, where she's lived, tastes in music and art. It's the starting point of any great relationship: the pleasures of discovering what's shared, and the chore of deciding if the differences are tolerable.

We talk and laugh our way into second glasses of wine. Meanwhile, I'm still processing what she's told me about her work. I feel an undercurrent of unease, a subtle shift in momentum. It's as if we're driving slightly too fast on that slick relationship road and the rear end of the car is ever so subtly beginning to fishtail as we round a corner.

Backstage

Alternative medical practices like naturopathy are here to stay.[3] But there are many vocal detractors. The skeptics quip,

> *Question:* What do you call "alternative medicine" that's been proven to work?
> *Answer:* "Medicine."

Are the alt-med skeptics being unfair? Snarky, maybe, but not really unfair given the evidence. Most objective reviews of research on alternative medical practices find little support.[4] The challenge is to locate the most promising treatments in an ocean of unsubstantiated claims and practices.

As advised in the internal conversation I'd had with my brain, I did some research on naturopathy. The field emerged in the early 1800s from folk remedies and practices in Europe. It's based on six principles, according to the American Association of Naturopathic Physicians (AANP). I found these to be totally reasonable maxims, including the following: promote the body's natural healing abilities, identify and treat the causes of illness, educate patients to encourage self-responsibility, and do no harm.[5]

The problem is that these principles don't address what should be naturopathy's central concern: to develop a viable theory and conduct systematic research to shore up its claims.

For example, *the* fundamental claim of naturopathy is that its treatments bring a patient's "vital energy" into "balance."[6] But these key concepts are undefined and unmeasurable. If *balanced vital energy* is really a thing, it would contradict what the laws of chemistry and physics tell us about bodies and energy.[7] Without evidence of its existence and a theory of its functioning, vital energy is a questionable foundation for an entire system of medical practice.

Naturopaths disagree, of course, and the websites of key organizations like the AANP and the Institute for Natural Medicine[8] publish spirited rebuttals to critical articles. Under careful scrutiny, however, these rebuttals can be shown to rely heavily on argument fallacies.

For example, the Wikipedia entry on naturopathy says it uses many "pseudoscientific practices," some of which are "outright quackery."[9] The AANP rebukes Wikipedia for being slow to correct falsehoods and too costly to sue for libel.[10] But those are misleading statements—"red herrings" as professional arguers would say. *Did* Wikipedia publish libelous falsehoods? No, their critical article on naturopathy musters scientific evidence to support its claims. The AANP article needed to refute Wikipedia with a theory that doesn't contradict the fields of physics, chemistry, and physiology, along with supportive research. It didn't, because such theory and research don't exist.[11]

Naturopathic websites do include links to supportive research. That's a good thing, of course, unless they purposely exclude unsupportive research—which they do—or unless some or all of the linked research they deem supportive is actually inconclusive—which is true for the relatively little pro-naturopathy research that's out there.

The very first linked article at the Institute for Natural Medicine's website illustrates the point.[12] Subjects at risk for cardiovascular disease were randomly assigned to one of two groups. One group had naturopathic treatments, the other didn't. Both received standard medical care. After a year, the naturopathic group showed modest benefits on several measures.

Alert readers see the design flaw. A proper control group would be treated exactly like the naturopathic group, except with a "sham" or placebo instead of naturopathy. The reason? Beneficial effects can follow *any* kind of extra therapeutic attention. A "treatment *vs.* non-treatment" design is weak and potentially misleading.

Another flaw in the study was that the naturopathic treatments included "health promotion counseling." This means subjects were encouraged to practice good diet and exercise habits. Since this intersects with conventional treatment, we can't know if the uniquely naturopathic treatments accounted for any unique benefits.

Marcia gave me an article on homeopathy, a treatment she used regularly. She'd had good success with it, even curing a patient's serious illness when nothing else worked. Experiences like this, plus the clinical evidence, put homeopathy at the top of her prescription arsenal.

Homeopathic products are easy to find, even in grocery stores and major pharmacy chains. They claim all sorts of applications, including restless leg syndrome, depression, digestive issues, hangovers, itchiness, the common cold, the flu, COVID, and loneliness. They are quite popular around the world, with over two hundred million users worldwide.[13]

The practice was invented in Germany in the 1790s by Samuel Hahnemann. It was based on subjective, self-administered "provings." By his procedure, he might eat a handful of hydrangea flowers and record what happens. The next day, he gets a headache and attributes it to the flowers. Then he applies the *Law of Like Cures Like*, inferring that hydrangea not only causes but also *cures* headaches. To prepare the cure, he applies the *Law of Infinitesimals*. According to it, the *smaller* the amount of hydrangea in the treatment, the *stronger* its curative effect.

Treatments are prepared to exacting specifications.[14] A typical recipe might look like this:

1. Prepare an ounce of pulverized hydrangea flower paste.

2. Put it into a hundred ounces of distilled water.

3. Dissolve and dilute the paste by shaking the container vigorously a hundred times.

4. Take one ounce of that liquid and discard the rest.

5. Return to Step 2. Repeat thirty times.

There are variants depending on the treatment. Six successive dilutions or a hundred; different numbers of shakes; pills of cornstarch or potions of water. There's no scientific rationale for any of the specifics. Just procedures without purpose, adhering to traditional rules handed down without research or testing. The "false precision" looks rigorous and scientific, but it rings hollow.[15]

What Hahnemann didn't know is that by the twelfth dilution, it's unlikely even a single hydrangea molecule remains. Yet this supposedly *maximizes* the strength. So the water somehow must remember a substance it no longer contains. It must also forget the millions of other substances that its molecules have contacted in their long lifetimes. There's no evidence that water does either, or a viable explanation for how it could.

The preparations' labels come with disclaimers like this one:

CLAIMS BASED ON TRADITIONAL HOMEOPATHIC PRACTICE, NOT ACCEPTED MEDICAL EVIDENCE. NOT FDA EVALUATED

In a statement about its position on homeopathy, the Food and Drug Administration warned the following:

> Products that have not been evaluated for safety and effectiveness may harm consumers who choose to treat serious diseases or conditions with such products, and consumers may be foregoing treatment with a medical product that has been scientifically proven to be safe and effective.[16]

No homeopathic treatment is FDA-approved. All remain unevaluated because they're not likely dangerous in and of themselves. No prescription is needed for a vial of water or a sugar pill.[17] But *safe* is a far cry from *effective*. Homeopathic proponents howl when called out on the fact that solid evidence of effectiveness is lacking.[18] But the fact remains.

An elderly James Randi, famed author and debunker, began a TED talk by downing an entire bottle of homeopathic sleeping pills—a massive overdose.[19] It had no effect on him. But let's think about this. Doesn't the law of infinitesimals mean that taking the tiniest portion of one pill should be more powerful than taking the whole bottle? Wouldn't the risk of overdose be from accidentally taking *too little*?

This is the paradoxical conclusion to which homeopathy's winding logical path leads. Not surprisingly, practitioners lack a consistent stance on overdosing.[20] The reason is simple. On one hand, consuming distilled water with nothing in it is totally benign—unless you overwhelm your kidneys with five gallons in a day. On the other hand, homeopathic remedies are claimed to be potent medicines that must be used according to explicit directions. How to capture this paradox on warning labels? Maybe like this:

> THE ACTIVE INGREDIENT IN THIS
> MEDICINE IS UNDETECTABLE.
> USE WITH CAUTION. TOO MUCH OR
> TOO LITTLE CAN HARM YOU,
> ALTHOUGH WE ARE NOT SURE WHICH.

Back to the article from Marcia.[21] Its author, Dana Ullman, claimed there are more than 150 supportive clinical studies. That's practically an infinitesimal amount of work for a field that's been around for centuries and which treats numerous medical issues for hundreds of millions of patients. In contrast, conventional medicine's registry of clinical trials logs several hundred thousand such studies *every year*.[22]

Ullman's article identified five references exemplifying homeopathy's high standards of clinical evidence. I'll summarize each of these. Quoted statements are from their authors.

1. There is a "lack of conclusive evidence" for most conditions; homeopathy "should not be substituted for proven therapies."

2. There is "insufficient evidence . . . that homeopathy is clearly efficacious for any of the observed clinical conditions."

3. "The evidence . . . is not sufficient to draw definitive conclusions . . . [because] most trials are of low methodological quality."

4. The author's own e-book, which reviewed material published elsewhere.

5. An experiment comparing homeopathic versus conventional treatments for vertigo. The two groups experienced equal improvement.

Ullman badly misrepresented numbers one through four by declaring they provided strong support for homeopathy.[23] What about number five?

There's a major problem: "Equal improvement" could occur for opposite reasons. For one, both treatments may have been effective. Alternatively, neither was effective. Neither being effective is a real possibility here because vertigo is so hard to treat.[24] But left untreated, it typically improves on its own in three to six weeks.[25] A third condition where a placebo was given would have clarified whether the treatments worked better than nothing. As it stands, it's just another inconclusive study.[26]

I wondered if my low opinion of the evidence in the article would have any impact on Marcia's faith in homeopathy. I didn't expect much. After all, she was a lot more invested in her practice than in me. But she claimed to be scientific. If she were serious, how would she rationalize prescribing pseudoscientific treatments?

My concern wasn't strictly for her patients. I also wondered whether she'd resent me as the bearer of bad news.

I wrote up a little summary for Marcia. Its scope was limited only to Ullman's article and its sources, concluding that they don't support homeopathy's claims.

Marcia replied in a detailed email. Many of her points are so commonly made by defenders of pseudoscience that I'm going to mention some of them here. She wrote:

> You absolutely convinced me that Ullman wrote a sloppy article, in all sorts of ways. . . . And if Ullman made the claim that scientific

studies had proven the efficacy of homeopathy, that's also unfortunate, since—you're right—it doesn't seem to have been adequately proven by traditional scientific methods.

So far, so good. Until the "but" butted in:

But some of the problem with proving the efficacy of . . . homeopathy might be the very use of those standard scientific methods . . . different tools might be needed.

Marcia didn't have a problem with scientific methods when she thought they supported homeopathy. So why are "different" tools now needed? Proponents and skeptics all agree that homeopathy claims *product X treats ailment Y*. Scientific methods are designed to test exactly these kinds of claims.

Marcia continued:

The tools we routinely use in studies might not be appropriate for proving the mechanism in homeopathy . . . [by which] the homeopathic remedy . . . is thought to stimulate the patient's "vital force" to restore balance.

I agree! Scientific tools are worthless for demonstrating an undefined, unobservable force. But we weren't talking about this. We were simply talking about whether it's been shown that any of the remedies actually work.

And then another point:

Are case studies alone proof of something's veracity? No, they're not. But they might still be worthy of attention, wonder, and exploration.

Absolutely true. But case studies are a form of anecdotal evidence and certainly *not* worthy of being the foundation for a system of medical practice. This led her to a popular trope about skeptics:

When skepticism rules the day to the point of crowding out curiosity, questions and exploration, there's very little room for learning, only reasserting one's original negative stance and getting to say, "I told you so."

But I *was* curious. I *was* open to the possibility that skeptics throw out the baby with the distilled bathwater. So I responded:

> I agree with you that if skepticism crowds out curiosity, not much new would be learned. But are skeptics setting such a high bar that real scientific innovation gets quashed? Science has shown again and again that it's open to wild new ideas, and it progresses despite its built-in skepticism.
>
> Science definitely resists accepting new ideas quickly, no matter how fervent their proponents. It waits. Sometimes it takes many years before the slow accumulation of evidence finally proves itself worthy.
>
> Homeopathy isn't being singled out for unfair treatment. Its core concept of "vital energy" is undefined and unmeasurable. Its research foundation is far from solid. Its remedies work no better than placebos. Short of meeting basic scientific standards, what's left is naïve faith in homeopathic authorities like Ullman and a pile of anecdotal evidence worth a penny a ton.

In the end, Marcia rationalized her faith with several strategies:

> When homeopathy fails tests, it's the research method's fault.
>
> When scientists are skeptical, it's because they lack curiosity about things outside their narrow paradigm.
>
> When evidence is of low quality, that's good enough for now.

Our relationship went into a full spinout, flipping over like a turtle, wheels in the air slowly turning to a stop.[27] Though our homeopathy exchange was totally cordial, within days, Marcia expressed her growing doubts about our potential. I couldn't blame her for feeling that way. But I'm glad we entertained the possibility of being together. You really don't know for sure if something works unless it's tested.

CHAPTER SIXTEEN

That's the Spirit!

Monsters are real, and ghosts are real too. They live inside us, and sometimes, they win.

—Stephen King[1]

FRONTSTAGE

They've been called specters, phantoms, spirits, spooks, and phantasms. They're all just ghosts, and my friend Ricky sees them.

For five years, we've played together in a band. Despite lots of rehearsals, weekend gigs, and time just hanging out, this is the first time this has come up. I knew every detail of his divorce settlement, his police record, his obsession with collards and fried okra, but not this.

On a perfect June Saturday in Charleston, South Carolina, Ricky and I were walking the promenade along the harbor. He grew up across the bridge from this old city and is giving me a history lesson. After passing three walking tour groups in the span of fifteen minutes, I casually observe, "There's a ton of these!"

"And that's just the ghost tours!" he points out.

My glib reply: "Wow, there must be a lot of gullible tourists."

Ricky jerks to a dramatic stop as though *he's* just seen a ghost.

"What's wrong?" I ask.

I turn back to face him, and in a hushed tone, he replies, "Barry, they're real." He sounds offended. "You seriously *don't* believe in ghosts?"

"I think they're really unlikely," a response I felt was both honest and diplomatic. I gesture for us to keep walking.

"But I've seen 'em," adds Ricky. "Lots of people have. How can you even say they're not likely?"

Should I just let it go? This is a classic *skeptic's dilemma*. It arises whenever one person doubts a claim that's near and dear to another's core beliefs. I don't want to start an argument, but I have to be honest. Also, Ricky's personal experience is probably going to trump any explanation I might offer. I dive in anyway.

"Okay. Two reasons. First, our brains are wired in a way that can mislead us. So we sometimes misinterpret things we think we've seen and heard and felt. Second, I've read a lot of serious ghost investigations, not like the ones on TV. The serious investigations have always found explanations that most people wouldn't know to look for."

Ricky knows he's in good company believing in ghosts. Especially here in the Charleston area. It doesn't take a psychic to predict what he'll say next: He *knows* what he saw.

"I don't know *what* I saw," he says.

Well, I had that wrong. And then he adds, "But I *know* what I saw!"

That seems to make a lot more sense to him than it did to me, but I don't press him on it.

There's a small business a couple of hours from home that's alleged to be haunted. The shop is called Vital Vapors, purveyor of all things vaping-related. Ben, it's owner, installed a security system with infrared cameras, microphones, motion detectors, door sensors, and digital storage.[2] The system recorded some weird things, and he wanted to have them documented for the record.

Ben hired Chip and his documentary film crew to tell the story. Then Chip emails me asking if I'd look at some of Ben's security camera videos. He explains that he'd make the film only if it includes a skeptical perspective. He offers to share a five-minute montage that Ben shared with him and that convinced Ben and others that the shop is haunted. I agree to look at it.

There are seven clips ranging from ten seconds to two minutes. Title screens direct attention to what the viewer is supposed to see and hear. They fall into five categories:

1. *Noises.* Specific sounds that stand out from background.
2. *Orbs.* Small blobs of light moving in a dimly lit room at various speeds and directions.
3. *Wisps.* Large, translucent waves of light moving through the image for several seconds.
4. *EVP.* Electronic voice phenomena: whispery words picked up by the cameras' mics.
5. *Flying objects.* Items fly off a shelf in the sales room when nobody's around.

These are way more interesting and important than any eyewitness testimony. I ask Chip to put me in touch with the owner. The next day, I've got Ben's email address, cell number, and permission to contact him.

I email Ben some basic questions. He says the video montage Chip gave me is just the tip of the iceberg. Not even the best stuff. He sends me eleven more clips, with more orbs, wispy figures, noises, and EVPs. I ask Ben if I can call, and he's eager to talk.

Ben's very open and trusting with me on the phone. He does most of the talking. I listen and comment respectfully as he tells of his experiences working after hours.

"When I'm in there alone at night, it's like there's a presence," he explains. "Sometimes I hear buzzes or clunks. I've heard whispers. Footsteps, too. I know how crazy this sounds, but I swear I can tell something's there watching me."

"How often do these things happen?" I ask him.

"Pretty much every night."

"Has anyone else experienced anything?"

"I have a guy who works part-time. I asked him if he noticed anything weird. He said 'Yeah, but not *that* weird. Maybe some funny sounds and smells.' He never paid much attention to it. But he doesn't work nights like I do."

"Anyone else?" I ask.

"Oh, sure. Some friends. Sometimes they hang out with me 'til three or four in the morning. They've felt things and heard things, too."

"Have you brought in anyone to investigate?" I leave it up to Ben to say who he feels is qualified to do this sort of thing.

"Yeah, a group of paranormal experts came a few times. They brought a spirit medium, too, and a bunch of electronic devices. People from my church came. Let's see ... Oh yeah, a heating and cooling guy checked the system to see if there's something in the air showing up on the cameras. He couldn't figure it, but everyone else verified something's going on that isn't good." Then he remembers, "I've also had sixteen spirit cleansings."

I assume that's some kind of exorcism for retail spaces. I have to wonder what kept them going after ten or fifteen failures.

Ben never once uses the word "ghost," preferring expressions like "it" and "whatever-it-is." Like my friend Ricky, he doesn't claim to know what it is, but he knows it's something. He *felt* it.

The Vital Vapors store sits on the inside corner of an L-shaped strip mall. It was empty for a year before Ben rented it. He now thinks it was vacant because the word was out that the place was creepy. He didn't find this out until he'd already committed to a six-month lease and set up shop.

Despite the strange goings-on, the business is thriving, and Ben's in the process of opening two more locations. He's overworked, stressed out, and sleep-deprived. He's also grieving the recent loss of his mother after a long illness. He mentions several times that her passing has added to his stress. He's a devout Christian who believes in God, prayer, angels, and demons. All this comes out in the hour-long chat, along with Ben's confession that he drinks too much and indulges virtually every night in the office.

I ask Ben if he's going to move the store when the lease is up. He gets emotional. I hear a struggle in his voice.

"This thing can't beat me," he declares. "I have to stay and fight."

Chip and his documentary film crew come to my campus to interview me. They set up two cameras and dramatic lighting in a darkened lecture hall downstairs from my office. I have plenty of material for them, but I'm leery because I don't know what others are saying in their interviews, or what spooky videos Chip might select for his final cut. What if I blather about orbs or silly ghost-hunting devices but they don't show them? I'll sound like a token skeptic, out of touch with paranormal reality. I try my best to cover all the bases as the cameras roll for more than an hour.

Being filmed is an ego-booster, I must admit. The first time was when I was hired as a "specialty extra" in the movie *Field of Dreams*. My scenes were filmed in Dubuque, Iowa, over the course of two nights. I was costumed in five scenes and got to chat with some of the stars during breaks. In one scene, it was just me and the director done up as Hasidic Jews chatting on a fake Boston sidewalk. It was his signature cameo, à la Alfred Hitchcock. A year later at the premiere, everyone cried at the sentimental ending. Not me. I cried because all my good scenes had been cut. I was reduced to a background smudge in a deli and a hand arranging meat inside a butcher shop window as Kevin Costner paused to look in. It was 97 degrees the day the butcher shop bit was filmed. There was no air-conditioning in the shop, and the meat moldered for many hours in the sunny window display as they did take after take. I deserved an Oscar for Best Suppressed Gag Reflex.

I was almost as disappointed seeing myself in the finished haunted vape shop documentary. I had two and a half minutes on screen, which was a lot more than usual for a token skeptic. But the result of all the editing was just a nicely lit professorial stereotype making vague skeptical noises, completely detached from the emotional experiences of the excited eyewitnesses.

That's showbiz.

The documentary had one surveillance video clip I'd not seen before. It was unlike any of the others. An open book that Ben identified as a Holy Bible rested on the far corner of his office desk, center frame. For a moment, nothing happened. Then the book started creeping on its own toward the edge of the desk. There was a three-second pause, and it suddenly flew off horizontally and could be heard landing somewhere out of sight.

If what the video seemed to show was real, then Ricky was right about ghosts, Ben has every reason to be disturbed, and I'll need to burn my skeptic's card and start walking toward the light.

Backstage

Ricky's belief in ghosts is not uncommon. He may not even be in the minority. National polls show that a third to over half of American adults believe ghosts probably or definitely exist.[3] In Great Britain, it's about 40 percent.[4]

TV programming and other media reflect these beliefs and help to fuel them. In the last twenty years, TV networks have aired dozens of popular ghost-related reality series in the United States, Canada, Great Britain, and elsewhere.[5] That doesn't even include print and social media, one-off "specials," documentaries, and feature-length films supposedly based on real events.

Ben was inspired by a popular "reality" TV series about ghost hunting. So was the paranormal investigation group whose help he enlisted. From them, Ben learned to hear words amidst the static of security camera audio recordings. He learned that orbs and wispy clouds are paranormal events; that "spirit boxes," electromagnetic field detectors, infrared cameras, and thermal imagers detect spirit phenomena with a sensitivity that exceeds human capabilities.

Ghost hunters taught Ben to trust his feelings. Ben learned to believe that what he *felt* at the misty boundaries of his senses was what *really was* happening. The fact that his videos also captured objects jumping off desks and shelves was pure paranormal gold.

TV ghost hunters are good at entertaining but lousy at hunting ghosts. If I had to boil it down to a root problem in these shows, it's how they use suspense, soundtracks, emotional outbursts, and incessant commentary to convince viewers that lots of really bad evidence is nonetheless good.

The dismal failings of typical ghost-hunting methods are detailed elsewhere.[6] But every element of the vape shop haunting is eerily familiar to those of us who've spent an irrecoverable hour of our lives on the likes of *Unsolved Mysteries, Scariest Places on Earth, Fear, Ghost Hunters, MonsterQuest, Ghost Nation,* or *Kindred Spirits.*

Let's take a closer look at each category of evidence.[7]

Noises. Some of Ben's security cam videos picked up strange noises: clunks, whirrs, taps, and other ambient sounds. I asked if he'd ever put recording devices outside his shop. Investigators often find that noises outside have been mistaken as coming from inside, especially at night when things are quieter. Ben hadn't tried that.

Vital Vapors is in a bustling strip mall at a busy intersection. One wall abuts another store. The other walls and ceiling face outside. There are parking lots front and side, an access road behind, a sidewalk along the front, a rooftop and back wall with heating and cooling units, a nearby regional airport, and heavy street traffic. With all the bustle, a jackhammer across the street could sound like tapping on a wall. An animated conversation at a parking space by the side wall may sound like muffled voices in the next room. It's easy to misattribute sound sources—especially if you're primed to think ghosts are vaping in the next room.

Besides, leaping to a supernatural conclusion before ruling out natural possibilities is just not good investigating. Leaping means not recognizing what's been leapt over.

Orbs. Orbs are small, light gray blobs, often in motion, appearing live or on video recordings in darkened rooms. Ben's videos show dozens of them.[8] We can't prove they're not spirits, but they're identical to the known "blooming" effect caused by dust particles or insects close to the camera lens.[9] Infrared cameras illuminate the particles, which blossom into orbs due to being out of focus. It's easy to replicate and could have been tested in the vape shop with a second camera: A spirit orb flying around the room would show up on both cameras. A particle of dust close to one camera's lens would not show up on the other camera. Ben had multiple cameras to play with, but they were each installed in their own rooms.

Ben knew about the dust explanation. But *his* orbs weren't dust, he maintained, because they moved when the heating/cooling system was off. It's good that he tried that little experiment, but he didn't take it far enough. Air can move for reasons other than climate-controlled systems: airflow through the ductwork from a door opening or closing in another room; drafts from doorways and gaps in the suspended ceiling tiles; convection currents from temperature differences between the room air and exterior walls.

It's worth emphasizing that "blooming dust" is an actual explanation for the images. "Spirit orbs" doesn't explain anything about what they are and makes huge, baseless assumptions about what they might be.

Wisps. Several videos showed wisps moving across the screen. Typically, a wisp would span vertically across the video frame and move across the room over the course of a few seconds. There are several possibilities for what they might be. On the simpler side, bear in mind that this is a store where people exhale a lot of moist, wispy vapors. There could also be some cobwebs around the camera mount.

On the more technical side, Ben had motion detectors in his security system. They emit infrared radiation known to cause artifacts in video images. There's also a possibility of "lens flares" from light sources interacting with the lens system—a moving screensaver on a laptop, or car headlights shining into the front windows and through a crack above the office door. You'd need to rule out these possibilities before titling a clip "Full Figure Flying Overhead," as Ben labeled one of his segments.

EVP. Electronic voice phenomena are all the rage in the amateur ghost-hunting world. They're alleged spirit voices caught on recording devices that are set to maximum sensitivity and left running, usually while ghost hunters mill around or yell questions into empty rooms, hoping for replies. Rarely is anything heard in person. However, listening to the recordings at maximum volume, sounds interpreted as voices are sometimes heard above the static.

In one of Ben's videos, the title slide directs the viewer to listen for the words "miss you." When I cranked up the volume, I could hear a hissing sound that *slightly* stood out from all the background static. I didn't hear "miss you" or any other words. In another video, the title screen leads viewers to expect to hear three whispered words. Again, the static was loud and clear, but there were no distinct sounds of any kind rising above the background.

The words that Ben and others heard were a type of audio mirage, a combination of *expectancy effects* and *auditory pareidolia*.[10] First, the title slides on Ben's clips inform viewers what words they should expect to hear. Then, pareidolia—the brain's tendency to find meaning in random stimuli—completes the job. The effect is powerful and surprising if you're not aware it's happening.

The other video with an EVP actually did sound like human speech. I could hear the word "inside" softly but plainly spoken. Unfortunately, nothing in the video rules out the possibility that it was exactly what it sounded like: a word spoken by a living person within range of a mic, perhaps unbeknownst to Ben. That's the conclusion requiring the shortest leap.

Flying Objects. There's a scene in the movie *Ghost* where the main character tries to kick a can on a subway platform. His foot passes through it, and he flops onto his back with an audible thud. You see, he can't yet interact with earthly objects until he learns to "focus." Eventually, he figures it out and kicks the can.

The *Ghost* writers tried to create a world where invisible spirits make visible things happen in our world. But the subway scene was beyond ridiculous. If the ghost didn't yet know how to interact with matter, how could he walk on the platform? Why would gravity pull him down? How could he land with a thud? Yes, this is just a movie, but the same kinds of questions can be asked in "real" ghost sightings. Don't expect to get any good answers.

Bottles, not cans, were the flying objects in Ben's video, as captured in broad daylight by the salesroom cam. He's behind the counter, then exits left toward the office. Twelve seconds later, four small bottles seem to leap off a store shelf on their own.

A closer look at the video reveals what really happened. Just as the little bottles took flight, there was an audible "*ker-chunk*," and the whole shelf section jerked down a fraction of an inch. Bottles were tightly packed, and most likely, those that fell were knocked off the front by those behind them. No ghostly hand required. But what caused the shelf to shift?

Still photos from the shop's website showed the glass shelving system up close. Shelves rested on a common type of notched bracket. If a shelf isn't nestled properly into its bracket, or a bracket into its notch, a small bump or vibration can cause the heavy shelf—and everything on it—to suddenly drop by a quarter inch or so. In this case, the vibration may have been supplied by loud music with a thumping bass line that could be clearly heard coming on in the background a few seconds before the shelf snapped down.

The paranormal investigator's job is to come up with reasonable, non-supernatural explanations. In Vital Vapors, every bit of ghostly evidence favors the normal explanation over the paranormal. A believer could still argue that a ghost clicked the shelf into place, or that the throbbing music was an elaborate EVP. In that case, I'd point out that until we know what a ghost is, explanations that summon *any* kind of unseen, unknown force are equally viable: the owner's telekinetic power, coordinated cockroaches, a giant invisible rabbit. All are as viable as the "whatever-it-is" in the vape shop.

The Flying Bible. This was different from the other video segments. There was nothing ambiguous about it. There was a book. It moved. It flew off the desk.

With no obvious explanation, it's perhaps tempting to leap to a supernatural cause. But I'll still advocate against leaping and for keeping an open mind about alternative explanations.

Magic tricks rely not only on the proverbial "smoke and mirrors." Some also come with strings attached. My first suspicion was that this was a simple string trick masquerading as a miracle. I couldn't prove it, but I could try re-creating it. I did, and it took only half an hour to set up and film a better-lit, higher-resolution version at my kitchen island.

My equipment consisted of an iPhone on a tripod, a dictionary, some duct tape, and a string. For atmosphere, Diana Krall's version of *I Don't Stand a Ghost of a Chance With You* played softly in the background. I'm seen from behind working at my laptop, the soon-to-be-flying book laid open on the far corner of the countertop. It moves, slightly. I gasp but shrug it off and go back to work. Fifteen seconds pass, and the book hitches toward the edge of the counter. My head jerks up, and I utter an expletive. Suddenly, the book leaps off the edge of the granite countertop. I turn, face the camera wide-eyed, and ask, "Did you see that?!"[11]

Of course "you" did. But what you couldn't see is that I'd duct-taped a string to the inside cover of the book, ran it around the counter, and held it in my right hand out of the camera's view. A little tug on the string and the book moved a little. A big tug and it took flight—very much like in

Ben's video. It would have been even easier to create the same effect if I'd have been off-camera.

This is more than a mere alternative explanation. Unlike any of the other videos, it implies purposeful deception on Ben's part—*if*, that is, the string method was used instead of the ghost method.

I called Ben and gingerly broached the subject. "How did you discover this sequence?"

"I was just reviewing the previous night's recordings as usual," he told me.

So I asked, "Is there any chance that somebody else might have been playing a little joke on you?"

"Nobody else was there that night," he promised. "And the video was time-stamped."

"Were you there?"

"Yeah. I was working in the front. I heard the bible hit the floor. But I also have two other videos where things moved in the office. I can find those for you," he offered.

"Sure, that would be great." I continued, "Hey, I have to say that I'm pretty skeptical about the flying bible clip. It was really easy to reproduce something just like it in my kitchen using string. I can show you my video."

He was firm. "All I know is what I've seen and heard and what's on the videos."

Ben never shared the other moving-object videos, and I never followed up, feeling that I'd seen enough at that point.

The extraordinary claims of the vape shop haunting are sorely lacking in extraordinary evidence. Sure, ghosts *may* exist. We don't know how they could, or what they'd be made of, or how they'd interact with our world. But if they do exist, not knowing these things doesn't mean they're unknowable. For now, the evidence strongly favors interesting but non-paranormal explanations that don't need to invoke unknown forces or energies.

I'm in no position to know whether Ben's judgment was skewed by his extreme stress, or his drinking, or his religious beliefs. It's possible, but my main interest was the paranormal claims and not why he might have been prone to believe them, or promote them, or manufacture them.

When I first saw the flying bible clip, I felt betrayed. At that moment, I believed Ben concocted the whole story to whip up publicity and bring in customers. But now I think it's probably more complicated than that. He may actually believe that the place is haunted, and that he needed to tweak some of the evidence to convince others. I'll never know, but I don't really need to.

What's troubling is that in my several conversations and emails with Ben, I never detected the slightest hint of insincerity. He struck me as a really good man. Intelligent, hard-working, and humble. And that's how I choose to think of him now.

My friend Ricky passed away in Charleston a few years ago. As so often happens with long-term bandmates, after ten years, our artistic differences became personal differences. I didn't have contact with him for the last five years of his life. But he's made my short list of people who've passed on that I'd still love to communicate with some more, if only there were a way. If he were to pay me a visit, I'd be thoroughly delighted to tell him to his face—or whatever he shows me—that he was right about ghosts after all.

CHAPTER SEVENTEEN

Numbers Game

A dozen, a gross, and a score
Plus three times the square root of four
Divided by seven
Plus five times eleven
Is nine squared and not a bit more.
—Leigh Mercer[1]

FRONTSTAGE

I'm cruising up I-81 North. Late January, light traffic. Skirting the edges of snowstorms, it's cozy-warm in the heated driver's seat. My hastily planned road trip to New York and New England is off to a smooth start. I'll be seeing a handful of friends, some old stomping grounds, and my parents' gravesite. I haven't been up there since my mom's funeral, already five years ago this month.

I like long drives. I pass the time with a rotation of podcasts, audiobooks, music, and daydreams. On a trip like this back to my home turf, I also spend a lot of time thinking nostalgic thoughts.

Other thoughts are numerical. Averaging 65 mph. Slowdowns ahead on I-78 and I-95. My 2014 Audi A4 rates 29 mpg but is getting 30.5. Heard it's $16 for the GW Bridge toll. Heading for 204th St. Haven't seen my NYC friend in, what, four years? Four and a half?

There's more.

The fifteen-gallon tank ranges 435 miles. Using 90 percent before refilling leaves a 43.5-mile cushion. Prices are around $4 per gallon, so $54 to top off the tank. First coffee was twelve ounces instead of sixteen, so I'll take only one bathroom stop about 55 percent of the way there, at exit 264.

Some numbers jump out at me from my dashboard. The trip meter hits 111,222,333 miles. GPS says it's 702 miles total and a 7:02 p.m. arrival. At some point, the clock shows 11:11.

I turn on a new episode of the *Radiolab* podcast, and the numbers in my head grow fewer and farther between. In Maryland, the odometer hits 75,000. That's hard to ignore. Big round numbers were fun milestones in my old cars—the same way it was fun to hit twenty years old and then twenty-five. Now the round numbers signify wear and tear and that it takes some work to keep us both running.

I stop for gas near New Market, Virginia, and check my phone for messages. I have an email soliciting an article for an online news outlet called *The Conversation*. They want me to think about numbers, too. They're asking for eight hundred words on the subject of "Twosday," coming up later this month. That's when the calendar strikes "**2/22/22**," which also happens to fall on a Tuesday. Get it? Twosday's on a Tuesday! One of *The Conversation*'s researchers noticed thousands of commemorative products for sale online and figured there must be something to this. I had fun working with their journalists and editors on some past projects, so I reply with a short note agreeing to write something for them next week. I put it on my mind's back burner and keep driving.[2]

Upon my arrival up north, the days and destinations roll by, filled with warm reunions and cold walks. Old haunts are visited, old memories are triggered. On the morning of the seventh day, I visit my parents' final resting place, a cemetery in Natick, Massachusetts. It's freezing, overcast, quiet, and breezeless. The central footpath from the street is plowed, but I have to crunch through some shin-deep snow to get over to the grave. I hadn't seen the double headstone since my mom's name was added. I just stand there a minute, my mind as still as the air. Inexplicably, I have an urge to talk out loud to them. So I say a few words about our relationships and regrets. I tell them about my daughter, their only grandchild. I mention that they're regularly in my dreams, where nothing bad ever happens with them.

Wrapping things up, I tell them, "I'm living my best life." Then, reassuringly, "These days, these years. . . . Sure, a few bumps. But I've been incredibly lucky. Thanks."

No part of me believes any part of them hears me, but it feels comforting and normal to be talking out loud through the blanket of snow, into the ears of the past.

I drive over to the next town, Framingham. I grew up here. My first nine years, we lived in an apartment in an old three-story house that was broken up into six small units. I'd not been inside since the day in 1965 when we moved to the other side of town. I've driven by the house now and then over the years when I visited the area. I've always fantasized about knocking on the door and asking to look around. This time, I do it.

I pull into the empty driveway and approach the side door. It looks like nobody's around, but I knock anyway.

"Who is it?" responds a woman's small voice.

I'm not used to speaking through closed doors, so I'm not quite prepared to answer this obvious question. But I try.

"Just a visitor. I used to live here. I haven't been back in a very long time. I was wondering if I can take a peek inside?"

I know how creepy this might sound to a suspicious person. Or to any rational person. Whatever social skill I need to ease that suspicion, I've never been more aware of lacking it than I am at this moment.

The door cracks open a couple of inches. It's dimly lit inside, and I can't make out anything. The lady's voice squeezes out through the gap.

"Sure—if you don't mind that I'm smoking a joint!"

Maybe being high has made her more trusting. I chuckle and say, "Not a problem!"

The heavy door opens wide into the kitchen. She motions me in and greets me with a big, open smile—this five-foot-nothing, overly trusting, stocky, middle-aged, stoned woman with no front teeth.

"I'm Barry," I tell her. "What's your name?"

"Maria."

"Hi, Maria. Thank you *so* much for this!"

I take a panoramic mental picture of the kitchen and dining area. I'm instantly transported back to childhood and have trouble finding my grown-up words.

"Lived here long?"

"Eleven years," she says. I notice the hands on the clock read 11:11. It's high up over the kitchen table, right where my family's old kitchen clock hung. "I like this place," she says. "I like the neighborhood."

I want her to know that I really did live here, too, and have no nefarious motives. So the first thing I do after introductions is point toward the wall at the back left corner and say, "Wasn't there a door over on that wall that opened to the basement stairway? What happened?"

"Oh, they took that out years ago." She explains, "Nobody can use the basement anymore."

Dang. I'd hoped to get down there, too. I loved spending time in the cellar. My apartment's small, dank corner of the house's underbelly was where I used my dad's work bench, played with his hand tools, and gleefully disassembled household objects to see how they worked. Never mind my persistent inability to put them back together again without leaving behind extra parts.

Maria animatedly talks about changes in the neighborhood and the town since she's lived here. Nearly everything she says is positive. I'm still feeling overwhelmed to be standing here. Nothing has changed in this kitchen. It's the same cupboards and the same double-basin sink where my mom bathed me as an infant. The range and refrigerator are in the same places. Her outdated kitchen table and chairs might as well have been ours—the scene of the crime where my mom served up her infamous mashed squash, a most memorable and awful side dish.

Maria walks me through the hallway to the living room. Her TV is where ours was, against the interior wall by the fireplace.

"It doesn't work," she says, pointing at the pristine hearth.

"It never did," I inform her.

Her couch is across the room, its back to the wall under the front windows, just like ours was. That was the hunting blind from which my brother and I, kneeling with our elbows atop the backrest, surveilled the neighborhood.

We walk through my parents' bedroom, then through the door connecting to the bedroom my brother and I shared. There was only one bed in there. It was on my side, in the same corner of the room where I dreamt of monsters and baseball games.

The loop complete, we're back into the kitchen. The thought of snapping phone pictures feels invasive, so I'm consciously trying to burn every visual detail into my brain—the cupboard hardware, the baseboards, the paint colors, the flooring. Memories are flowing—images of mundane family scenes, of moving about these very rooms so long ago. My voice is quavering, and I'm close to tears. I focus on the uneven drywall above the stove where the nonworking exhaust fan used to be.

When it's time to say goodbye, I thank Maria effusively for opening her door to me. She hugs me.

This place. I went from being unconscious to conscious here. It incubated my curiosity. It fed me a snack after school. This is the place I'll likely never see again after I walk out that door.

I walk out that door.

I hit the road for home early the next morning. Before the sunrise, I'm already thinking about the "Twosday" article and what I might say about it. The first things that come to mind are other standout date patterns. Eleven years ago, there were two different *Ones*days—1/11/11 and 11/11/11. *Three*sday, *Four*sday, and *more*sdays are yet to come. Casting a wider net, we've had twelve days since 2001 with patterns like 01/01/01, 02/02/02, and so on.

I'm a little surprised to realize that I totally missed that the day before yesterday was 2/2/22. But I'm pleased to realize for the first time that if I'd been born thirty days sooner, my birthday would be 3/4/56. Thirty-five years later and it would be 4/3/21.

It's a *long* drive.

The article I'm going to write will have to bring in other kinds of numerical weirdness. I come up with three: the Lincoln–Kennedy coincidences, the practice of numerology, and the Bible Code. The first of these brings to mind a coin-collecting incident from my childhood.

I'm seated across the kitchen table from Tommy, a fellow ten-year-old coin collector. We're mostly talking pennies and nickels today, showing off our latest acquisitions. Tommy reaches into a shoebox and pulls out a small cloth bag held shut with a drawstring. He says he just got this as a gift. With hushed reverence, he unpuckers the top of the sack and upends it near the other coins we were examining. A clear plastic double-coin case clicks onto the table.

It's *so beautiful.*

It displays two gleaming uncirculated specimens: a copper Lincoln penny and a silver John F. Kennedy half-dollar. On the back is a white label obscuring parts of the coins. It contains some tiny print: a series of numerical "facts" linking the two murdered presidents.[3]

- 6   day of the week (Friday) of both assassinations
- 7   letters in Kennedy's & Lincoln's last names
- 8   successors' birth year (both "Johnsons"): Andrew 1808; Lyndon 1908
- 15 letters in assassins' names: John Wilkes Booth; Lee Harvey Oswald
- 21 letters in Kennedy's & Lincoln's sons' names: Patrick Bouvier Kennedy; William Wallace Lincoln
- 39 assassins' birth year: Booth 1839; Oswald 1939
- 60 year elected—Lincoln 1860; Kennedy 1960

I have an odd feeling in my tummy, like I'm hearing a ghost story. These connections *have* to be more than just a coincidence. But beyond that feeling, I really don't know how to process it.

Many years later, I learn that the Lincoln–Kennedy list is a type of offshoot of *numerology*—the practice of searching for occult significance in anything relating to numbers.[4] A more popular version calculates a "Destiny Number" from any name.[5] Knowing this number, you can look up information about your personality, behavioral tendencies, and, of course, your destiny. It works like the magic decoder rings they used to put in cereal boxes. I'll illustrate with my name.

First, each letter's number is given by this coding scheme:

A:1	B:2	C:3	D:4	E:5	F:6	G:7	H:8	I:9
J:1	K:2	L:3	M:4	N:5	O:6	P:7	Q:8	R:9
S:1	T:2	U:3	V:4	W:5	X:6	Y:7	Z:8	

My first name—B, A, R, R, Y—gets 2, 1, 9, 9, 7.
Next, add these numbers: 2 + 1 + 9 + 9 + 7 = **28**.
Third, repeat for middle and last names: **22** and **36**.
Fourth, add those results: 28 + 22 + 36 = **86**.
Finally, add the digits from that result, and repeat until only a single digit remains: 8 + 6 = **14**, 1 + 4 = **5**.

This is my Destiny Number. Any reputable numerology book or website will have descriptions for all nine, mostly innocuous compliments. My "5" means, among other things, I'm both enthusiastic and easily bored.[6] *Nailed it!* I was enthusiastic several paragraphs back. Now, I'm bored. So let's move on to a more interesting illustration.

The Bible Code was a best-selling book in the 1990s.[7] Its author, Michael Drosnin, claimed that the Bible contains coded messages, including prophecies that later came true.

How did Drosnin crack God's ancient code?

He used a form of numerology. The process is tedious but not very complicated. Working with the Hebrew version of the Old Testament, he arranged the entire text into rows and columns.[8] Just like in a word search puzzle, the words can run horizontally, vertically, or diagonally, and they can be read forward or backward, up or down. The difference is that in a regular word search puzzle, the correct answers are arranged by a human puzzle maker. In the Old Testament puzzle, the words were supposedly arranged by God Himself.

Numbers come into play in the form of "skip codes." These were simple instructions for skipping characters instead of reading only adjacent ones. Suppose you find these letters embedded in a row:

Y E M C H A R N T Y T O W U K F V I P N J D O M R E Q L K Z

A skip code of "1" creates strings out of every other letter. In this case, we'd first extract

Y M H R T T W K V P J O R Q K

followed by this other new row:

E C A N Y O U F I N D M E L Z

Nothing much jumps out from the first extraction. But embedded in the second line, we find the phrase "CAN YOU FIND ME." But we needn't stop there with the skipping method. We can also scan a book's entire text, skipping two letters at a time, three at a time, and so on, with no upper limit. A skip code can also be compounded as in "3 right, 12 up."[9] All of these manipulations are handled by a computer program, and a large number of them were attempted in the Bible Code. Most importantly, the computer program sifts through all the extracted letter strings in search of familiar words, names, and phrases.

Drosnin's discoveries were quite remarkable. For example, "Yitzhak Rabin" and "assassin that will assassinate" were found intersecting one another in the decoded text. Israeli President Rabin was assassinated in 1995.[10] Also, "Kennedy" appeared near "Dallas" where the US president was assassinated. "Hitler" was found close to "Nazi." Other decoded text suggests that an "atomic holocaust" is imminent.

Much time and effort were spent on the development of a screenplay for the book. Alas, *The Bible Code* book failed to prophesize that *The Bible Code* movie would never come to pass.

I've been following my weather app as I make my way home and watching the temperature fall all morning in western Pennsylvania. It's 31 degrees and drizzling. Black ice on the roads, white knuckles on the steering wheel. I make it through the area before it gets too slick. Luckily, the temperature rises as my route arcs due south. It's raining but otherwise easy driving for a long, long way.

After dark, I'm climbing the mountain pass in western North Carolina where I-26 reaches up nearly a mile. It starts snowing. More good luck: I make it through before it accumulates on the road. So much fortunate timing today. It's good to be a "5" on the numerology scale.

At about 9 p.m. I pull into my parking space back home. I pause a moment to reflect on all that I've seen and done on this trip. Random thoughts. The beadboard cupboard doors in the kitchen of my childhood home. The sweep of so many decades. The arcs of regular lives.

That my days are numbered only increases their value.

BACKSTAGE

Twosday, 2/22/22. Why is this so attention-grabbing? Is there anything to it beyond the coincidence of repetition? Is it more than just a date marker? More than the boldly emblazoned T-shirts and coffee mugs? Is there anything *there* besides the collective regret of those who spent $30 on some very specific merch that was discounted to $4.95 on Wednesday, February 23?

No. I can say this with a lot of confidence.

There's no covert, numerical espionage at play. Numbers are as innocent as babies. They signify quantities, as in "five gallons of gas." They label things, as in "Interstate 5." And they count things as in "I'm going home in five days." Any significance attached to numbers beyond these kinds of uses is on *us*, not them.

That is to say, numbers are assigned all kinds of meanings when the context is social rather than mathematical. Our brains evolved a fantastic capacity to find patterns and connections.[11] Doing so once meant the difference between survival and death. Being able to recognize different paw prints in the dirt, for example, or weather patterns in the skies was crucial to living long enough to sire offspring and perpetuate one's species.

Even if food, clothing, or shelter isn't at stake, it's rewarding to notice a pattern such as a familiar face or song.[12,13,14] Finding one, the brain zaps its synapses with a pleasant little shot of dopamine, motivating us to keep finding more patterns.[15]

When a number seems to stand out from the background and call attention to itself, this is one kind of *apophenia*: finding meaningful connections between unrelated things.[16,17] Nonnumerical examples include conspiracy theories, seeing faces in clouds, star constellations, and the belief that breaking a mirror causes bad luck.[18] Apophenia is intensified when you're consciously looking for patterns or primed to notice certain ones. Once you've been told that "11:11" has spiritual significance,[19] you're far more likely to notice it—as I did on both my car's digital clock and the analog kitchen clock in my old apartment.

Snipe-hunting for number patterns—whether in presidential assassinations or the Old Testament—is the stock and trade of numerology.[20] It masquerades as scientific by operating with numbers. But it's not a branch of mathematics. It falls squarely within the realm of *divination*, alongside palmistry, tea leaves, and crystal balls.

Numerology hangs around because it satisfies all of apophenia's urges. It marches before our eyes an endless parade of patterns that make us feel things: control over life's chaos; the thrill of exploration and discovery; a sense of empowerment; satisfaction and wholeness from the feeling of learning more about ourselves and our destinies.

These feelings are misleading. There are plenty of legitimate reasons in the world to experience discovery or empowerment, but consumers beware: Numerology isn't one of them.

The rhetoric of numerology sounds authoritative, as when it confidently assigns personality descriptions to Destiny Numbers and Destiny Numbers to people. But it's rooted in speculation and spiritualism, not in valid observations and tests.[21] It's harmless fun to follow the procedures and find your result. But as a source of empowerment or self-discovery, it leaves everything to be desired.

A little clarification is in order before we revisit the Lincoln–Kennedy coincidences. Let's distinguish two types of coincidences: "just-a" and "more-than." To say something is "*just a coincidence*," it literally means "co-occurrence." You embody the coincidence of a particular height and weight simply because they co-occur. "Just a coincidence" is utterly unremarkable.

To suggest that something is *not* just a coincidence points to the second type: the "*more-than-coincidence.*" It's a co-occurrence overlaid with the amazing. It implies that some unknown cause is behind the co-occurrence. It feels like more than coincidence, for example, that the same woman would give birth to three girls, spaced three years apart, all sharing the same birthday.[22]

Whether or not something seems more than coincidence is a poor indicator of whether or not there's a deeper cause or whether only chance was at work. Sometimes there isn't enough information about events to be able to tell one way or the other. Oftentimes, there is, but it may take some digging.

All my examples, including the Lincoln–Kennedy facts, are of the more-than-coincidence variety. As the list unfolds, each new line ratchets up the wow factor. You may feel a tingle in your spine the first time you see it. It certainly made ten-year-old me feel something, and it made the memory unforgettable.

In this case, the hidden cause behind the more-than-coincidence feeling is simple: human intervention. The Lincoln–Kennedy line items were deliberately culled from a much larger pool of potential co-occurrences *only because* they're among a few with interesting patterns.

Mentalist Derrin Brown has a video demonstrating another way this works.[23] He looks into the camera declaring that he'll try to flip a coin into a glass bowl and get ten heads in a row. This would certainly qualify as seeming to be more than coincidence. A close-up of the bowl appears on split screen. He begins the flips, each time announcing the result—heads every time—and the count of consecutive heads. After the tenth in a row, he looks into the camera, thanks the viewing audience, and walks off camera. Spooky!

Probability theory (definitely *not* numerology) calculates the chances of this happening as about one in a thousand.[24] There are ways to cheat at coin flipping, but his coin appeared to be fair, and the video wasn't edited between flips. So . . . how?

Simple: the magic of video. He recorded the flips *until* he got ten heads in a row, then edited out all the previous flips. Probability theory predicts that obtaining ten consecutive heads will take 2,046 flips, give or take.[25] Brown took about eight seconds per flip, so the taping session would have been expected to run about four and a half hours. For a professional illusionist, a half-day shoot for a spectacular video segment is well worth the time.[26]

The Lincoln–Kennedy numbers work the same way. A few interesting co-occurrences are drawn from a very large set of possibilities. Hide the boring ones (for instance, those whose hometowns had different numbers of letters in them), and those that remain are framed in a way that gives them far more significance than they deserve.

What's in the mostly hidden set of Lincoln–Kennedy possibilities? The list I showed earlier offers some clues. There were numbers for days of the week, years of events, and letter counts. Multiple people were mentioned: presidents, assassins, presidents' children, and vice presidents. Two kinds of events were listed: assassinations and elections. Already it's evident that there are many potential combinations of words and numbers. The original pool likely included many more, such as place names and other relatives and associates. Try to imagine a giant grid filled with all of these numerical facts, referring to all sorts of people, places, and events. It makes for thousands of co-occurrences. That a dozen or two of them are interesting isn't surprising. Leaving all the dull ones off the list is no different than Derrin Brown deleting all of his unsuccessful coin flips.

The Bible Code found entire words and phrases using numerical skip patterns in the text. Once again, the *more-than-coincidence* illusion comes from emphasizing certain interesting letter strings and word combinations and ignoring vast numbers of uninteresting ones.

How large was Drosnin's set of possibilities? It's almost inconceivable. He used the Hebrew Torah, a version of the Old Testament of the Bible that contains several hundred thousand characters. But other factors exponentially increased the chances of finding interesting coincidences:

- Allow letter strings to run in any direction. This multiplies the chances of finding recognizable words.
- Try different vowels. There are no vowels in the Torah, so a given letter string could be interpreted multiple ways.
- Choose different dimensions for grids. This generates different letter strings for the pool.
- Try dozens, or hundreds, or thousands of different skip codes.[27]
- Retrofit to known events. For example, keep Kennedy–Dallas, but toss out Kennedy–Cleveland and Kennedy–Waffle House.

These methods generate *billions* of letter strings searchable by word-recognition software.[28] This casts serious doubts on Drosnin's claims. Undeterred, his response was to double down and issue a bold challenge:

> When my critics find a message about the assassination of a prime minister encrypted in *Moby Dick*, I'll believe them. [*Newsweek*, 6/9/1997]

Mathematician Brendan McKay accepted the challenge and succeeded.[29] He also retrofitted "prophecies" for many other prominent deaths—Lincoln's and Kennedy's included.

Despite his method getting crushed under the weight of *Moby Dick* and other unholy texts, Drosnin continued to endorse his findings up until his own death in 2020.

I've always enjoyed working and playing with numbers. But if it hadn't been for professional baseball player Carl Michael "Yaz" Yastrzemski, I might have gone the numerology route instead of studying regular old mathematics and statistics.

Yaz was my boyhood hero, a star player for the Boston Red Sox. Back then, I believe there was a law requiring Massachusetts parents to naturalize their offspring into "Red Sox Nation" citizenship. In this cult, the ball players are god-men, Fenway Park is mecca, the official language is Bostonian ("My fathah's cah has a wicked pissah Red Sahks bumpah stickah."), and *malaise* was the prevailing sentiment because our accursed team had not won a championship since 1918.

Baseball cards were the coin of the realm. We'd occasionally buy a fresh pack at Dom's Market for a nickel—five cards plus a stick of bubble gum. These cards were the raw material of an underground economy run by little boys. Because some cards were more valuable than others—Yaz's above all—commodities needed to circulate. Bartering and gambling were the prime movers. Bartering was a straightforward way to acquire a valuable card: Thirty-five average players' cards for one Hall of Famer's rookie card, for example. Gambling involved different card-throwing games such as Topsies, Knock-downs, Stand-ups, Nearsies, and Leansies. The games were more wide-open than trading and a way of acquiring large quantities of average cards for future bartering—but only if playing "for keepsies."

When not haggling and gambling, we studied the backs of the cards. For there were printed the player's career highlights and statistics. These were only very basic stats, but they gave us plenty of fodder. We argued about which players were best, who should be traded or sent down to the minors, and which players from other teams we'd admit to wishing played for our side.

These charts were my first exposure to cumulative data analysis, summary statistics, and trend interpretation. Maybe it was because my earliest contact with stats was pure fun that I later developed a deeper appreciation for numbers. Knowing a thing or two about statistics can be life-changing. I highly recommend it.

Yaz played major league ball for twenty-three years, all with the Red Sox. He retired in 1983 and still co-holds the Major League record for the longest tenure with a single team. In 1989, the Sox retired his uniform number. No other Red Sox player will ever again wear the great Yaz's number eight.

What does numerology tell us about Carl Michael Yastrzemski? We can start by calculating his Destiny Number with the formula I gave earlier. We get 16 for Carl, 33 for Michael, and 49 for Yastrzemski. We add those three numbers and get 98. We add those two numbers (9 + 8) to get 17. And miracle of miracles, adding 1 + 7 gives us Yaz's legendary uniform number, 8, as his Destiny Number.

Just a coincidence?[30]

CHAPTER EIGHTEEN

Circling the Square

Three conspiracy theorists walk into a bar. You can't tell me that's just a coincidence!

—Anonymous internet meme

We've come to the final chapter. It's longer than any before it, though probably shorter than it should have been. There's a lot going on. The Frontstage section is a conversation with someone close to me who bought into the extraordinary claims of a popular conspiracy theory. How this could happen to her is a vitally important, multi-faceted topic because it can happen to practically anyone.

To break things up a bit, we'll pop back and forth between conversations and explanations, Frontstage and Backstage. Later in the chapter when we return to the conversation, we find the believer reacting to some facts that contradict her conspiracy beliefs and that she'd previously either brushed off or never heard.

The Backstage materials are a fitting capstone to this book. They draw from research areas ranging from the psychology of emotion to structural engineering. They help to explain key aspects of a widely publicized event—the World Trade Center collapse—and why some people still see it so very differently from others. As usual, the endnotes suggest where to find more information or to help someone you care about escape the grip of false and sometimes harmful conspiratorial beliefs.

Although the World Trade Center tragedy was a generation ago, it persists in our collective psyche. We'll get to it in a bit. The rest of this story is very recent.

FRONTSTAGE

I pull up to Tina's apartment and text her, "I'm parked outside." Momentarily she appears at her door, struggling with a large ... thingamajig. It's sweltering out, and her face is shiny with sweat. She flashes me a smile as she passes the front of my car. Ah, now I can see she's wrestling with a computer monitor. Actually, a double monitor—two regular flatscreens bracketed side by side on a stand. The bracket is loose, and the screens swing around wildly.

She crosses the parking area to a dumpster, reaches the opening, and hikes up the wonky hardware to shoulder level. For a second, it's balanced on the lip of the dumpster, but it lists back toward Tina as though fighting its inevitable fate. As it starts to fall on her, she shifts her feet and catches it awkwardly. Again, Tina hoists it up to the edge, this time managing to shove it over. It lands inside with a sonorous *clunk*.

She heads back in my direction, and I exit my car. Our chuckles make for a jiggly hug.

"Smooth moves over there at the dumpster," I tease.

"Listen. It didn't land on my head, and I didn't fall down. I'll take that as a victory, thank you. But sorry you had to witness it, Uncle B."

"Teen! Forget I said anything. But the slapstick was a nice bonus. Anyway, how *are* you?"

"Great!" She's cheerful despite having some struggles with adulting. But lately things seem to be going pretty well.

Tina is the daughter of close old friends, my "Uncle" title merely honorary. I've known her since her infancy thirty years ago, and I've always been her "Uncle Barry." We stay in touch through emails and letters and try to meet up a couple of times a year, coinciding with my visiting her mom and dad. Today, we're heading over to her favorite restaurant, China First. We arrive, place orders at the counter, seat ourselves, and dive into conversation.

"So what was that about—when I pulled up at your place? The monitors you dumped looked almost new."

"Yeah, really new," she admits. "A gift from Jon, my ex-boo."

"Ahh, I get it." I assume the screen-dump is symbolic of the boyfriend-dump. But she continues, hushed and leaning in closer.

"They ... were watching me ... through the camera," she says, adding, "and yeah, I know how crazy that sounds."

I ask the obvious question: "Who's 'they'? Who's watching?" Another reasonable question would be whether her desktop monitors even had a built-in camera, but I let that one go.

"It may be the government watching. Big Brother." Is this her dry sense of humor, or is she serious? I play along.

"Why?"

"Weird stuff's happening," she explains.

"Like what?" I ask, trying not to sound judgy or dismissive.

"Like, spam emails with clues in them, or that mention something personal. It's like they can watch me in my apartment. There's 'malware' that can do that, you know. Oh, and I get calls with blocked numbers. I answer, and they don't say anything. They just breathe."

She goes silent, visibly upset. I let a minute pass and ask, "Is there more?"

"Well, yeah. Lots. They broke into my place. More than once. Things are moved around, like they're looking for something. I put in a little 'nanny cam,' but something always happens to it, and it doesn't work right just when I need it to."

I have so many questions, but I don't want to swamp her or get her defenses up. I say only, "That sounds scary. Why do you think it's the government?"

Normally wide-open in conversations with me, Tina measures her words.

"Not the *whole* government. Maybe some kind of law enforcement. State Police? FBI? CIA?" Before I can ask what they could possibly want with her, she explains: "Something happened. This guy Daniel. He's friends with Jon, the guy I broke up with. Daniel's a bad dude, into drugs and guns. He has it in his head that I owe him, like, five hundred dollars or something."

I'm scared for her. Who are these people she's involved with?

"Daniel won't take 'no' for an answer," she continues. "But he won't hurt me because Jon still cares about me, and they're buddies since they were kids. So instead of shooting my kneecaps, I think Daniel's punishing me other ways."

"Like spying on you?" I ask.

"And sending me coded messages. Or getting someone else to. There're just too many weird things. They can't be coincidences. Here. Look at this. Came yesterday." She finds an email on her phone, opens it, and hands it to me. I read it aloud:

> This is you final notification. You are in arrears with your federal taxes and must remit full payment in the amount of $652.78 before March 1st to avoid further penalites.

I read it as written—"pen-a-lites"—but look up at Tina and say, "They mean 'penalties'?" She nods.

> If you fail to repond ["re-POND?" I ask rhetorically], IRS agents will come to your residence and collect payment. This can be avoided if you're account is cleared by March 1st.

I can't keep myself from grammar policing. I interrupt my reading again: "It says Y–O–U, apostrophe, R–E instead of Y–O–U–R. Jeez!" Continuing on,

> Negligence on your part will result in late fees, punitive fines, and garnishment of your wages.
> We strongly advise you to call this toll-free number immediately to arrange for payment: (888) 733-1205. All major credit cards are accepted.

"This isn't from the IRS," I say, handing back her phone. "They don't contact people through email. And I'm pretty sure they know how to use a spell-checker."

"I know!" says Tina emphatically. "And anyhow, I paid my taxes."

"So why take this seriously?"

She holds up her phone, pointing to the screen. "It looks like a form letter, right? Where they fill in the amount I supposedly owe? But read the first word in each paragraph."

"Okay." I lean in, squint, and read aloud. "'This' . . . 'IRS' . . . 'Negligence' . . . 'Any.'"

"And the first letters of those words?"

I look again. "T . . . I . . . N . . . . A. Okay, 'TINA.' I admit, that's pretty weird. But it could be a coincidence, right?"

"Sure," she replies, "*if* that was the only thing. Look at the phone number. Last four digits. Recognize anything?"

"One-two-oh-five. No."

"It's a date," she points out.

"Oh, okay. Twelve-five. Your birthday. But . . ."

"There's so much more," she interrupts. "Every day there's stuff like this. They're messing with my electronics. My alarm clock goes off randomly sometimes. And every night, if I wake up, the time is '1-2-3-4' or '5-4-3' or '2-2-2' or some other pattern.

"Then there's this weird hum all the time from the other side of my bedroom wall. And my friends have been telling me about stuff people are saying behind my back—that I'm supposedly doing bad things to people. It's all lies! But I wouldn't put it past Daniel to start rumors about me, or to make something up and report me to the police anonymously. Like I said: bad dude."

Tina's getting upset now, and I don't know how to make it better. I say, "I'm really sorry this is all happening." But I feel almost as helpless as she does—except I don't believe most of it.

"Thanks, Uncle B. Do you think I'm crazy?"

"Well, yeah Teen," I say, smiling. "I've always known this about you." She laughs, and I jump at the chance to derail the conversation by saying, "Let's talk about your job, and try to enjoy these eggrolls."

And so we do.

Two seasons pass before I see Tina again. We meet for lunch at China First as usual, finding a cozy table by the large front window. Outside, it's snowing lightly. A clicking baseboard heater next to our feet buffers us from the cold radiating off the frosted pane.

Her occasional newsy communications have been positive since the last time we met. A job change, a little dating, a foster kitten. Nothing about being surveilled or harassed, though I never asked her directly. Now that I'm sitting across from her, I'm dying to know what's happened with all that. Awaiting our order, I consider how to raise the subject in a way that gives her an easy out if she'd rather not talk about it.

I decide on, "That thing with the IRS. Did it ever amount to anything?"

"Nah," she shakes her head, "and the other stuff tapered off. Maybe I overreacted a little. What-evs. It's behind me."

I breathe a loud sigh of relief.

"I'm sorry if I worried you, Uncle B. I blame the friends I was with—and myself for choosing them. All of them were sort of, you know, paranoid. And into drugs. Last fall, I backed away from them and started making some new friends."

"That's great! And what are *they* like?" I ask.

"I've been hanging a lot with Olivia and Eli. They're a couple. Really smart. They like to talk about science and politics and philosophy."

"Do you usually agree with them about stuff?"

"Usually. Not always. I try to keep an open mind," Tina replies. Then, in what seems like an odd digression, she asks, "Do you remember when I was in first grade and 9/11 happened? And my parents asked you to pick me up early at school?"

"Sure I do. They were stuck at work. My teaching was cancelled so they asked me to get you. What do you remember about it?"

"I was only six," she said, gazing distantly out the window, "but I remember *everything*."

"Yeah, me too."

She continues, "Teachers coming in and out of our class whispering to each other. Ms. Walker announcing there was an accident in New York, rolling the TV cart to the front of the room. Seeing the fire and all the smoke. I wasn't too scared because Ms. Walker wasn't making a big deal about it. It just seemed like the usual tragic stuff you always see on the news.

"Then things changed when the second building was hit. Ms. Walker looked worried. Teachers were talking to each other in the hallway. I was looking at the other kids to figure out what I was supposed to think. They were all over the place—some were goofing around, some were terrified. Ms. Walker turned off the TV, and parents started coming to pick up their kids, one by one. That was actually the scariest part. By the time you came, half the kids were gone. It felt like families were huddling up for the end of the world."

"Remember when we got to my house?" I asked.

"Yeah. You tried to distract me with mac and cheese for lunch. And you talked to me calmly while the TV kept showing the towers falling over and over again. You kept saying we're safe 'cuz New York's far away."

"I was torn, Teen. You wanted to know what was really happening. Should I have told you what was going on? Or should I have shielded you from it 'til your parents picked you up? I tried to be honest with you, while filtering the truth for your innocent brain."

She quickly adds, "Or what they wanted you to *believe* was the truth."

"What do you mean?" I ask.

"My friends Olivia and Eli call themselves '9/11 Truthers.' I've learned some things from them about the World Trade Center bombing." Reading my expression, she disclaims, "I *don't* believe everything they say. They think 9/11 was a plot to rule the world. They think powerful people wanted to rally Americans behind a huge war. But seriously? The CIA, and the 'military industrial complex,' coordinating a plan to attack America? With George 'they-misunderestimate-me' Bush pulling the strings? I don't *think* so."

"What do you agree with them about?"

"That the towers were blown up by explosives and a lot of people were involved. But I don't think we know who did it."

I pause another beat, then ask, "You don't believe in the *other* conspiracy?"

"There's another one?" she asks, incredulously.

"Yeah—where nineteen members of al-Qaeda *conspired* with Osama bin Laden to fly planes into American buildings?"

Tina shrugs. "Maybe, but the Truthers say that's the lie put out by the people who are *really* behind it all."

"But what if it's the Truthers who're caught up in some *other* group's false conspiracy? Know what I mean? Like, maybe ETs from Proxima Centauri b teleported explosives into the Twin Towers and started the Bush–CIA rumors."

She smiles at that. "I know you're an astronomy geek, Uncle, but you just made that up. There's no evidence . . ."

"Not true," I interrupt. "It's an Earth-like planet in a nearby solar system. There's no evidence it *doesn't* have intelligent life. If it does, fifty-fifty chances they're more advanced than us. We've sent out radio signals for almost a century bragging about our resources. We're ripe for the plucking. Maybe with 9/11 they were testing our defenses. Connect the dots—like the Truthers do."

"Okay. Sure." But I know she doesn't believe me for a second. "Look, I've read a lot by legit engineers who say it *had* to be explosives. Plus it's obvious in the videos."

"What do those engineers say?"

"Wait a sec." She pulls up a notes app on her phone, opens a page from the index, and reads to me:

1. The impacts of the planes alone couldn't bring the buildings down.

2. Burning jet fuel can't melt steel beams.

3. Right before the collapse started, videos show the explosions blowing out the sides, timed to explode lower and lower to collapse the building.

4. The buildings collapsed into their own footprints, looking exactly like a controlled demolition.

5. Explosives could have been hidden in replacement ceiling tiles in the offices and detonated wirelessly.

6. Chemical traces of explosives were found in the debris.

7. Pools of molten steel burned for weeks underneath the debris pile, far hotter than any kind of building fire.

8. Nanothermite from the explosive devices could have fueled the burning.

Tina asks, "Should I go on?"
"No, that's enough. I . . . I . . . I think you've convinced me!"
"Really?" She asks, looking excited.
I nod my head yes but say, "Are you kidding me? Of course not. Ha! Got you again!" She rolls her eyes. "Come on. Let's get started on these eggrolls, and I'll tell you what's really behind all this. And no, it's not the Proxima Centaurians."

Tina's eyes go wide as she looks over the delicious lunch—eggrolls, mu shu tofu, and egg fu young. She says, "Please, *please* don't tell me it was the Chinese."

BACKSTAGE

It wasn't the Chinese. And given how competently Tina was living her life overall, I don't think she's crazy. I will say that, while we all latch onto false beliefs sometimes, Tina's concerns about being spied on and harassed were tinged with some paranoia. It probably didn't rise to the level of a clinical disorder,[1] but I was concerned because it troubled her at times and encroached on some aspects of her daily life. I was happy she got past it.

Cognitive psychologist Steven Pinker argues that a certain amount of paranoia is normal, handed down to us through evolutionary roots.[2] It stems from an imbalance between the costs and benefits of overreacting to perceived threats. Back when the world was a more dangerous place, surviving long enough to pass down your genes favored those who overreacted. You're hunting for food and hear a rustle in the grass. If you assume "Poisonous snake!" and run away, it costs you no more than a little exerted energy if you're wrong. But if you underreact and decide "Oh, it's nothing," then being wrong may cost your life and your place in the gene pool. Pinker calls this *evolved constructive paranoia*. You're rewarded, genetically, for being a scaredy-cat, even if real threats are rare.

A little paranoia can also have some benefits. Tina was on her own, living with fear and anxiety brought on by the perceived threats of the "bad dude." Whether or not she was generally prone to paranoid thoughts, she was trying to manage some scary uncertainties. So she started looking for patterns. Finding some, it gave her a degree of comfort to believe she'd figured out what was going on. To the outsider, it was plain to see that she was probably weaving stories out of unrelated threads. But to her, the resulting designs were real and insidious—but at least within her grasp.

Paranoid thoughts are usually about "them" threatening "me." Others, for reasons known or unknown, want to harm me. In contrast, the "them" in conspiracy theories is always a coordinated group or coalition. Their target may be one person or many. We, the observers, may or may not believe we're being personally targeted.

Similar to Pinker's concept of evolved constructive paranoia, the idea behind *adaptive conspiracism* is that conspiracy beliefs fixate us on potential dangers in our social environments. It's generally better to muster defenses against a conquering coalition that may not exist than to fail to prepare for one that does.[3]

A conspiracy theory is a many-tentacled, hyperaggressive, unstoppable myth. It is the Kraken of beliefs. Such theories have thrived in the United States for at least the last two hundred years.[4] Followers can number in the thousands, or even millions, despite having little or no good evidence to support their faith.

Nonbelievers find conspiracy theories ridiculous. They're baffled as to why anyone would subscribe to them. Believers see their validity as self-evident and often assume nonbelievers are willfully ignorant.

Even as a child, some conspiracy theories struck me as nonsense. Others are notable for their popularity, sheer absurdity, or destructiveness. I'll list some examples here, originating in each of the last eight decades.

Holocaust Denial: 1940s
Official story: In a World War II genocide, Hitler directed his forces to murder approximately six million Jews in concentration camps.
Conspiracy theory: This so-called Holocaust was greatly exaggerated, if not a complete hoax.

Flat Earth, 1950s
Official story: Earth is roughly spherical in shape.
Conspiracy theories: NASA and others suppress evidence that the Earth is a flat disc and fake evidence that it's spherical.

President Kennedy Assassination, 1960s
Official story: JFK was killed by a lone gunman.
Conspiracy theories: The assassination was carried out by multiple gunmen with support from the CIA, and/or Mafia, KGB, US government, or others.

US Moon Landings, 1970s
Official story: NASA's Apollo program successfully landed people on the moon for the first time.
Conspiracy theories: NASA faked the landings to avoid humiliation by Russia and to ensure future funding. It enlisted Hollywood filmmakers to simulate the events.

Roswell Incident, 1980s
Official story: Debris from a crashed military surveillance balloon was recovered in Roswell, New Mexico, in 1947.
Conspiracy theories: Many years later, several authors popularized a claim that the military was covering up evidence of an alien saucer crash.

Chemtrails, 1990s
Official story: Jet contrails in the sky are caused by water vapor in engine exhaust and low temperatures at high altitudes.
Conspiracy theories: Contrails are chemical or biological agents, evidence of government programs to control the weather, test weapons, or sicken people.

9/11, 2000s
Official story: Members of al-Qaeda, a militant, pro-Islamic group, flew planes into American buildings.
Conspiracy theories: The World Trade Center collapse was a controlled demolition; the US government was behind it all.[5]

QAnon, 2010s
Official story: The QAnon conspiracy theory is ludicrous, having emerged from people role-playing on the 4Chan website.
Conspiracy theory: Anonymous insider "Q" portrayed President Trump as secretly combating an international ring of Satanic, child-abusing cannibals, among them Democrats in the US Congress.

Presidential Election, 2020
Official story: President Biden beat former President Trump in the general election by over seven million votes and by seventy-four in the Electoral College.
Conspiracy theory: The election was stolen from Trump via a combination of malfeasances, including unauthorized voters, hacked voting machines, and corrupt election workers.

COVID-19 Pandemic, 2020s
Official story: SARS-CoV-2 (severe acute respiratory syndrome coronavirus 2) originated in Wuhan, China, and spread worldwide. It killed close to seven million people, over one million in the United States.
Conspiracy theories: The virus was a bioweapon released accidentally; a tool for population control unleashed by Microsoft founder Bill Gates; an unintended consequence of 5G cellular networks. The Centers for Disease Control and the Department of Health and Human Services exaggerate morbidity and mortality rates.

There are many more, and much has been written about conspiracy theories from every angle imaginable. Two recent books on the subject include Mick West's *Escaping the Rabbit Hole* and Michael Shermer's

Conspiracy.[6] These are noteworthy because, taken together, they cover the conditions for the emergence and persistence of conspiracy theories and offer information to help someone escape them.

Conspiracy theories aren't all cut from the same cloth, but they do tend to share some common features. For one, they're carried on the backs of *beliefs*. Holding a belief means mentally associating a particular set of properties with a specific thing. For example, Tina associated the properties "inside job" and "controlled demolition" with a specific event—the World Trade Center collapse.

A belief can range from weak and inconsequential to powerful and life-altering. Several factors contribute to its strength. First, are its associations numerous and solid? Think of the associations as stilts protecting a house from being undermined by floodwater.

Second, is the belief espoused by others who are respected and socially close? It's validating and affirming when one's beliefs are shared with others whose opinions we value.

Third, does the belief work in concert with other related beliefs? A conspiracy theory's web-like structure strengthens its central claims and protects it from inconvenient realities.

Fourth, how central is the belief to the holder's identity? Does it define the person? Would it damage their sense of who they are if the belief were to prove false? Would it mean losing social ties?

Beliefs form instinctively and can do so whether or not what's "in here" (pointing at my head) conforms with what's "out there" (pointing out my window). Like optical illusions, false beliefs can arise due to the ways that perfectly normal brains process information. I'll illustrate this using an extended analogy. Normally, I'd rather lay out the concept directly. But in this case, I think—no, I *believe*—the analogy makes the point more powerfully.

Start with this simple pattern of dots:

• •

• •

Most people see it as a set of four dots forming a square. It's hard not to because we mentally connect the separate elements into a single coherent unit:

Our brains take the path of least resistance. Instead of dealing with the dots as separate objects, it reduces them to a simple shape.

The problem with this is, like stars in constellations, the dots may not actually be connected to each other so simply, if at all. The mind can mislead. Even worse, once it sees a square, it takes extra effort for the mind to see it differently. The belief is sticky and hard to shake off.

Seeing the pattern as a square also reduces the motivation to see it any other way. But if the dots aren't really part of a square, then thinking outside the box, so to speak, becomes the only way to see it differently. For example, you may inadvertently have "squared the circle"—an unsolvable challenge for mathematics and a metaphor for attempting the impossible.[7] But it's no problem for human perception, which can square circles, and circle squares, with ease.

The four dots also could have been extracted from a triangle, suggesting the awkward but apt coinage "triangling the circle."

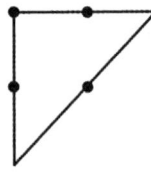

The point is that sometimes there's simply not enough information to know the "true" shape from the dots you're sampling and *you don't know there's not enough information*. It might have been a different story had there been eight bits of information instead of only four, as in these figures:

But doubling the number of data points still can't ensure we'll trace the shapes correctly. Definitely not in this case because I made the eight-dot circle using two "squares" and the triangle is really the original square shape plus a diagonal line. So neither pattern is as it might first appear.

There's always going to be room for alternative interpretations when "connecting the dots." We need to try to keep our minds ajar. We may not have all the information needed for a sound judgment, and we wouldn't want to find ourselves holding onto a false belief just because it seemed the most obvious.

When I talk with people who are skeptical of a conspiracy theory, they'll often say two things about believers that go something like this:

1. "I don't understand how they can fall for that stuff."
2. "They feel special thinking they know secrets and insider information."

The second point partly answers the first: We believe things that make us feel certain ways. Let's consider how this and a few other factors come into play.

Culture. Tina was raised in an American culture that begets and grooms conspiracy theories. It's not uncommon to hear about bad people hatching elaborate plots to harm or exploit common people. In modern society, they're usually spawned in the primordial soup of social networks, then propagated everywhere by mass media.

Institutions. Over decades or even centuries of cultural evolution, we've fashioned a variety of very large institutions in areas that include education, family, science, entertainment, politics, sports, religion, and the economy. In turn, we are shaped by the institutions we've created. To flesh out their understanding of 9/11, people can selectively draw information from multiple institutional spheres such as political (why the terrorists' countries hate America), religious (the perpetrators' pro-Islamic and anti-Christian beliefs), scientific (why the towers really collapsed), and mass media (news outlets imposing their own points of view).

Social Interaction. Tina didn't sleuth her way through dusty archives to reach independent, evidence-based conclusions about 9/11. She took an easier route. Her information had already been selected, prepackaged, and delivered by her friends. As with most social influences, Tina was unlikely aware of the full extent to which her newly held beliefs originated in the minds of others. They felt more like well-reasoned conclusions to her based on valid evidence, even while knowing her friends were the catalyst.

Perception. Perceptions are interpretations of signals from the senses. As with the four-dot analogy, one may perceive the same thing more than one way. While Tina watched and rewatched videos of the towers collapsing, she was directed to focus on certain details, and to deemphasize others. This led her to perceive them from a new and very specific perspective.

Cognition. Thinking about one's perceptions—consciously or otherwise—introduces a wide range of potential influences based on memories, personalities, desires, and expectations. *Motivated cognition*, for example, is a process that distorts perceptions and beliefs to better align with one's preferences. Tina was motivated to be accepted by a new group of friends. Adopting their conspiracy beliefs helped. She learned to see things as they did by allowing herself to be selectively exposed to supportive information and shielded from anything or anyone that might challenge it.

Emotion. The events that generate conspiracy theories tend to evoke strong feelings, and we frequently use emotions to guide our judgments. Going with our gut reactions is a good example.[8] This includes some of the most important decisions we ever make, such as friendships, careers, financial moves, spiritual beliefs, and life partners. Beliefs which have been amplified by emotions resist change. Passionately held beliefs are more likely to evoke action, such as when self-identity is at stake.[9] Tina was working through transitions in her personal life, including the end of a romance, a job change, moving on from a set of friends, and seeking new friendships. This was the emotional context in which she learned about 9/11 conspiracy theories, their fraught imagery and intrigue perhaps serving as distractions from her anxieties.

BACK TO FRONTSTAGE

Suddenly, Tina bolts upright, and her eyes twinkle. I've seen this look before. She's about to do her hilarious impression of a social media influencer. She takes some duck-lipped selfies and food pics, then raises an amply stuffed mu shu pancake off the plate. Still looking into her phone camera, she moves the food nearer her mouth to frame a close-up. She side-eyes the drippy little burrito and bursts forth with a cheerful endorsement to her adoring audience: "China First's mu shu tofu. Mmmm, a savory delight! Dine-in, take-out, or delivery. When I have a hankerin' for something yummy, I always consider China . . . First! See what I did there?"

I almost spit out my jasmine tea.

When I finally compose myself, she takes us back to the main discussion thread. "Uncle B, go ahead. Tell me why it wasn't a controlled demolition on 9/11."[10]

Tina's known for years about my course on pseudoscience and paranormal beliefs, but we've never talked about conspiracy theories, and I didn't used to talk about them in class. But over the last few years, I've done a lot of course prep on the topic. Most of it focuses on conspiracy theories that emerged since 2000. So I'm well-prepared to talk with her about 9/11.

"I don't know anything for sure," I began, hoping not to sound too teacherly, "but I've heard all the arguments you listed. Each has another side to it. It's like connect the dots." I grab a pencil from my jacket pocket, a napkin from the dispenser, and give a quick lecture on how the mind tends to gravitate toward wrong solutions.

"Okay, I get your point Uncle . . . *in theory*. But the rigged explosives weren't a connect-the-dots game. The way the towers fell meant only *one* thing, not lots of different things."

"There I disagree with you, Teen. So much of what happened took place out of anyone's view. There were actually *big* gaps between the dots. And most people aren't structural engineers with the appropriate knowledge to offer judgments."

"Maybe," Tina acknowledges, "but I still think some dots are obvious patterns."

"True." She's listening more intently than most of my students ever do in class. "But a pattern isn't the thing itself. It's more what it looks like from a certain angle. The dots I showed may suggest squares and circles, but they're just dots. But when someone sounding like an authority says you're being lied to and offers an interesting alternative, it doesn't mean it's the only alternative or the correct one. And once we see the lines drawn *their* way, it's hard to see them any other way."

"But Uncle, that applies to you, too. You don't think it was explosives. That's your 'square.' Have you even tried to see the 'circle'?"

"Sure, and it's really not hard to see it both ways if you try. It helps if you don't have any personal stake in who's right or wrong. But suppose both sides are biased. Is there any hope of ever getting to the truth?"

She ponders for a microsecond and replies, "Evidence?"

"Yes! We have to find out what people with real expertise have to say about it. But this gets tricky. Often, people with little real expertise are good at coming off like they know things. Now, let's go back to the list of reasons you read that says why it had to be a controlled demolition."

"Okay." She pokes her phone screen a few times. "Let's see. One: The impacts of the planes alone couldn't bring the buildings down." She looks up at me.

"Okay, sure, it would be good to know if that's true. But what is it supposed to be evidence *of*?"

She thinks about that for a moment and answers, "Yeah. No. I guess it's not. It sounds more like it's a conclusion that needs its own evidence."

"Exactly! So, is there evidence that planes *can't* destroy towers?"

"Not that I've seen," she says. "But honestly, I've focused more on whether the explosives could have done that."

"Okay. But it's still super important to know if the planes alone could bring down the towers. Let's start with some basic facts I've learned about this claim: The towers' framework was steel girders, right? Jet fuel burns at under 1,500 degrees Fahrenheit, which is about 1,200 degrees too low to melt the steel."

"Exactly!" says Tina, thinking she's just scored some points. "Burning jet fuel couldn't have melted the frames and brought down the buildings."

"*But*," I continue, "nobody's claiming the steel beams melted." I wait for that sink in.

She answers, "What? There's this group, 'Architects & Engineers for 9/11 Truth.' They say the beams couldn't melt but that the official accounts claim they do. Why would they say that?"

"I don't know," I answer, "and I won't try to guess. The question is, what do nearly all *other* engineers say about this on the record?"

"What do they say?" Tina asks.

"That the support beams didn't need to melt for the collapse to happen.[11] Some beams broke from the force of having jets crash into them.

Also, steel loses 50 percent of its strength when heated to only 1,200 degrees. So rather than melting, floor beams most likely sagged from the heat of the fires. Those sagging beams strained their riveted ties to the vertical beams. Eventually, some of those joints broke and started a domino effect. There's evidence of this breakage in the piles of rubble. You can actually see where the joints broke apart.

"But the joints didn't all have to break. Some of the sagging floor beams pulled the vertical beams inward. Picture a can of soda with some dents on the side. This also contributed to the collapse by weakening the ability to hold up the floors above. Once you put a side dent in the soda can, it's much easier to crush the whole can."

Tina ponders and says, "Okay, maybe that made a few floors fall. But the whole building, top to bottom? Remember point number three: 'Right before the collapse started, videos show dozens of explosions moving down the building ahead of the collapse, blowing out the sides.' When you watch, it's really clear that's what happened."

I ask Tina if she's watched many videos of controlled demolitions in other tall buildings. "Not really," she says.

"You should. They don't look anything like the Twin Towers' collapse. In real demolitions, they only put explosives at a few strategic locations—first and foremost, the ground floor. Gravity does the rest of the work. The weight of the higher floors crushes the floors below.[12]

"In the World Trade Center videos, what look like explosions blowing out the sides are huge puffs of air. A floor above falls onto the floor below it, and air and debris are explosively forced out the sides. But that stuff shooting out is a *result* of the collapsing floors, not the cause."

While Tina's processing this, I ask, "What's next on the list?"

"'Four: The buildings collapsed into their own footprints, looking exactly like a controlled demolition.'"

"Oh yeah. Well, that's just not true." I Google "ground zero" and show her the first photo that comes up. "Here's the site after the collapse. Looks more like an *un*-controlled demolition, right? The mess went way outside the footprints. But anyway, controlled demolitions don't always come straight down. Some videos show them purposely making the building fall sideways into an empty lot. They do this by blowing out only part of a lower floor. That tips the collapsing building in a planned direction—like making a tree fall this way or that by chopping a notch into one side of the trunk.

"Rewatch the videos, but this time, see how the part of the buildings above each plane crash tipped a little toward the weakened sides where the planes hit. Then gravity took over, crushing floor after floor, pancaking them down one on top of the other."

I didn't bring up a similar but partial collapse that happened in London, but I'll mention it here. A gas explosion on the eighteenth floor of a twenty-four-story apartment building blew out load-bearing walls at one corner. This left the corners of the six floors above it unsupported. They fell into the damaged unit, the momentum causing a progressive collapse down the entire corner, all the way to the ground. No explosives required.[13]

"Alright, Uncle B," Tina chimes. "I'm skeptical, but I promise I'll think about that some more. Send me some links?"

"Sure. Now what about the rest of the list? The explosives?"

She checks it again and says, "Number five: explosives in the ceiling tiles. Okay, I can guess what you're going to say: 'Just speculation, and too many people needing to keep a big secret . . .'"

"Yup," I agree. "Ceiling tile bombs. Seems a little far-fetched without any evidence for them."

She adds, "Well, there are lots of different ways explosives can be planted."

So I ask, "Is there any real evidence for *any* of them? Was a single person ever caught? Was a single unexploded device ever found?"

"No, I don't think so. But there *was* evidence of explosive materials after the collapse. That's numbers six, seven, and eight on my list." She looks at her phone again. "Chemical traces, plus pools of molten burning steel that couldn't happen from the fires, probably fueled by nanothermite from the explosives."

These are all popular elements of the conspiracy theory, so I'm ready. "Tina, it would take a month of lunches to unpack all that. I'll just say a couple more things.

"First of all, most of the main people in the engineers and architects group you mentioned aren't structural engineers or demolition experts.

Three of the most vocal members are a software engineer, a physicist, and a theologian.[14] Sounds like a walked-into-a-bar joke, right? But they're basically just hobbyists, not trained investigators. They personally couldn't understand how planes alone could bring down the towers, so they wrongly assumed that planes couldn't. Simple as that.[15]

"Second, pools of molten steel would eventually have cooled to form giant steel ingots in the basement. Nothing like that was ever found.

"Third, the only indicator of thermite ever found was a trace amount of manganese. This later turned out to be misidentified, a common error. There just wasn't any evidence of explosives at all, "nano" or otherwise. But the group *needed* there to be evidence, so they never retracted their nanothermite claim. Also, none of the workers on the site, or chemical analysts, or structural engineers ever found anything indicating explosions. It was just a few guys promoting a false scenario under the guise of a scientific organization."

Tina took it all in with humility, but I knew she wasn't convinced. And that's okay for now.

Back to Backstage

"Just because you're paranoid doesn't mean they aren't after you." So wrote Joseph Heller in his best-selling novel *Catch-22*. Even in the midst of her struggle with paranoid beliefs, Tina seemed to understand this implicitly. But she couldn't quite let go of the idea that she was being spied on through devices planted in her home. If you've never had someone you care about caught in a web of paranoid thoughts, consider yourself lucky. It's hard to watch.

In a similar vein, *just because it's a conspiracy theory doesn't mean there isn't a conspiracy.* Real conspiracies happen, including the one where al-Qaeda hijackers crashed planes into the Twin Towers. But some conspiracy theories are far less plausible than others. While it's unlikely there was a second JFK assassin, the idea is not nearly as absurd as the idea that NASA has been promoting fake news of a round Earth.[16]

Fortunately, there are ways to separate truth from nonsense. Michael Shermer adapted Carl Sagan's general-purpose *baloney-detection kit* for use on conspiracy theories. Among other things, he recommends asking:

- Is the claimant legitimate?
- Do multiple sources of evidence converge?
- Have accepted rules of research and evidence been applied?
- Are alternative explanations being ruled out rather than ignored?[17]

One or more of these qualities is always lacking in a false conspiracy theory.

Still, in practice, it can be hard to tell. Most conspiracies, including the 9/11 Truthers', are buttressed by a rash of related claims about the perpetrators, their motives and methods, and minute details of the event and its timeline. Disprove one belief and there are plenty more to fall back on. Who has the time to apply baloney-detection tests to every single detail?

Crazy as it seems, some *make* the time. Michael Shermer, Mick West, Carl Sagan, and many others I cite in the endnotes have devoted decades of their lives to sifting through baloney. Paleontologist Stephen Jay Gould loathed debunking because it took time away from his real work. But debunk he did because, he wrote, "Skepticism is the agent of reason against organized irrationalism, and is therefore one of the keys to human social and civic decency."[18]

I'm not sure I'd paint skeptics as superheroes preserving all that is good in society. Most of us just like finding out the best possible explanations for extraordinary claims.

Curtain Call

I love Tina like a daughter. I want the best for her, and it hurts my heart to see her stressed out with paranoid thoughts or wasting her time watching disaster videos over and over under the tutelage of conspiracy theorists. But it could be worse. She had someone she trusted who could offer counterpoints to key conspiracy claims. And as far as I know, she wasn't tempted to join Awake Dating, a website where conspiracists can "meet someone who shares your 'socially inconvenient' understandings and truth-seeking ethos."[19] Anyway, I heard that Awake is an innocent-looking front organization for Big Cupid, a Russian cabal bent on destroying Western culture.

Sadly, Tina's had friendships end because she wouldn't buy into the QAnon conspiracy. Another friend of hers is certain the Illuminati are running the world. She knows people who think our national elections have been stolen and that software mogul Bill Gates put microchips into COVID vaccines.[20] She hasn't embraced any of those conspiracies. I'd even say that she's generally skeptical. But like all of us, when people she cares about believe something crazy, it seems a little less crazy.

It's okay to be uncertain, but that doesn't mean it's easy. If it helps, remember that feeling uncertain in a complex world is a sensible way to be. Uncertainty is a foot in the door, holding it open a crack while better information is gathered.

It's after two o'clock, and the lunch crowd has thinned to a handful. Outside, the snow flurries have stopped, the sun has broken through, and judging by the wet sidewalks, the temperature's gone up.

"A walk before you go?" asks Tina. We have a regular route that takes us a couple of blocks south to the town square. We typically stroll around it a few times, sometimes stopping into one or two of its shops before returning to the car near the restaurant.

"Sure!" I reply, and we zip up our coats, put on our gloves, wrap our scarves, and step outside.

Walking and talking is different from sitting across a table. You're not as locked into the other person and freer to be more wistful or blunt. We talk about the way the town transforms when spring comes. The sun, the warmth, the leaves returning, and the walkways bustling.

We take turns solving the world's problems in short monologues.

We talk about time passing and parents getting older. She more than compensates me for her lunch by telling me I've not aged a day since 1995.

The subject of 9/11 comes up again, and I decide to go for the bludgeon: "I'll be honest with you Tina. I hope you balance your intake of conspiracy stuff with some good skeptical sources. I really do."

Tina goes for the wistful: "Uncle, you know, the truth doesn't care what I believe. I'll connect the dots this way, or I'll connect them that way, and the world will keep on turning."

I think about that for a few paces. We pass the snow-dusted benches and dormant chessboard tables in the town square.

"You're so right, Tina. Either there were explosives or there weren't. What you or I believe doesn't change that. But there'll be other times when what you believe affects you and those around you a lot. The more important the issue, the more important it'll be for you to know how to get your hands on good information. It may be your health at stake, or your finances, or your relationships, or your happiness."

"I hear you, Uncle. I've heard everything you've told me today. Especially the point you made with the four dots and the patterns. And guess what?"

"What?" I ask.

She gestures toward the benches, statues, and the dry frozen fountain. "We just circled the square!"

She winks. I laugh. The Earth turns a little bit more.

EPILOGUE

Awe and wonder are irresistible pleasures. To fuel and deliver these addictive feelings, we build industries, institutions, organizations, and black markets. We engage as a culture with travel, art, music, film, religion, and psychotropic drugs. As individuals, we tickle our wonderment bones on the cheap by stargazing, walking in nature, curling up with a good book, or falling in love.

It's less common to acquire these kinds of feelings by exploring their causes. In fact, it's popular to believe the opposite—that studying extraordinary claims, and ultimately learning something from them, negates all the fun.

Obviously, I disagree.

At least *some* people out there find the process and results of digging beneath the surface of an extraordinary claim even more wonderful and awe-inspiring than simply taking it at face value. And doing so has the nice side benefit of getting closer to the truth about what's really happening. Not everyone needs to feel this way, but I wish it were more widely known that there are exciting alternatives to unquestioning acceptance.

I don't mean to imply you can't walk in both worlds. I may get some of my jollies from learning what's really going on with dowsing or UFO sightings, but I'm as vulnerable as the next person to the spine-tingling awesomeness of the night sky, a great museum, a hike in the forest, an amazing song, an all-encompassing novel, and love. Supernatural stories are fun. Supernatural beliefs are, too, but ought not be mistaken for the natural world—which is already quite super without the adornment of myths and fallacies.

ACKNOWLEDGMENTS

One morning, I had some free time before meeting my writer friend Kathy Sheldon for breakfast. I started jotting down topics and titles for articles I thought might be fun to work on. I had loads of material from years of teaching my Sociology of the Paranormal class to college undergrads, and it seemed like it would be fun to compile some of my first- and secondhand experiences related to those topics.

Twenty minutes later, I had over thirty titles. Only then did the thought occur to me that this could be a book that a lot of people might want to read. I had decades of irrelevant professional experience publishing articles and chapters on esoteric topics in my academic field. But this book *wanted* to be written and insisted I try.

I brought my list of chapter titles to breakfast and read them to Kathy. In addition to being a writer and editor, she was a key person at a small publishing house that used to be in our town of Asheville, North Carolina. I knew that I could trust her judgment about whether the titles sounded like a plausible basis for a book. Had she been lukewarm or less, this book would not have happened. But Kathy was enthusiastic, and that motivated me to get writing. Since that morning, her advice and continued friendship have filled the project's sails.

Few people have known me better or longer than Rose Garfinkle. She reads tons of books, and I trust her judgment. So what she said to me after reading a draft of an early chapter convinced me I might be onto something: "It reads like a great story, and then suddenly you realize, 'I'm learning stuff!'" That's exactly what I needed to hear at the time. Thank you for your encouragement, love, and support over the years.

What also kept me going was a coincidence almost worthy of its own chapter. I was meeting online monthly with one of those pandemic-

ACKNOWLEDGMENTS

spawned Zoom discussion groups. One of its members, Zoë Blaylock, is a writer (among other things) from San Diego. For reasons I still can't fathom, she thought I might fit into a small, eclectic, international writing group she'd recently organized. I knew I'd need lots of critical feedback on my writing, and this seemed like just the ticket. At that time, I had so far drafted only part of one chapter. But I accepted her invitation for a tryout the next day.

Everyone in that group was a published author, and everyone's ten-to-fifteen-minute readings that day blew me away. When my turn came around, I was fully intimidated. Fortunately, Zoom allowed me to keep my trembling hands off-camera while I steadied my voice and read my piece. They liked what they heard, at least enough to take a chance on me. Their generous feedback over the subsequent months forever changed my writing for the better. Thank you, Zoë, for inviting me into the group and for your many valuable comments. Equally heartfelt thanks to our group's other West Coasters, Shira Musicant and Eben Mishkin. And to our elegantly accented British members across the pond, Jon Higham and Neil Shaw. I hope we six will meet in person someday. I could never repay my debt to you, but dinner's on me anyway.

I discovered scientific skepticism in 1980. There are too many inspirational publications and authors to list, so I'll only mention some of the most influential. I've referenced the two leading skeptical journals throughout the book: *Skeptic Magazine* and *Skeptical Inquirer*. I love these publications. Key authors now living only through their writings are Carl Sagan, James Randi, and Martin Gardner. Still very much alive are the indefatigable Michael Shermer and Ben Radford. I won't speculate here on the unfortunate absence of women on this short list—a symptom of the demographics of the pool of publishing skeptics. But I'll gladly add Mary Roach and Sharon A. Hill to my list of recent influencers.

No surprise if you know me, but I kept a spreadsheet for those who read draft chapters and communicated reactions to me. Various relationships are involved here, ranging from the owner of an Airbnb in Massachusetts where I stayed a few nights, to dear friends, family members, and a partner-for-life. Your support has been amazing. You are the audience I try to please. Alphabetically, you are Lisa Baugh, Jo

ACKNOWLEDGMENTS

Ann Beard, Lyn Benjamin, Cynthia Cobb, Natasha Dale, Rose Garfinkle, Neal Jones, Eli Lamport, Paul LeValley, Tess Markovsky, Lynn Patters, Rachel Rosner, Jeri Senor, Kari Sickenberger, Wendy Wiberg, and Barbara Wright. Thank you for reading drafts and laughing at me in mostly the right places.

I had the benefit of doing chapter readings with two different live audiences while working on the book. It's one thing when friends and family laugh at your jokes and claim to have learned something from your work. It's quite another to face the judgment of strangers. The fact that both talks went well helped fuel my motivation to see the project through. Thank you to the organizers and members of the Asheville Science Tavern, and the National Capitol Area Skeptics. It was a blast!

Finally, Jeri Senor. You are a supernatural experience. I didn't even know you when I started this project and now look at it—and us! What I've learned from you about love and compassion has made me see things in new ways, both for the betterment of my writing and myself. Now, my wonderful forever partner, thanks to you, everything is a beautiful work in progress.

NOTES

Chapter 1: Boy's Nose Amputated by Cousin
1. Roald Dahl, *The Minpins* (London: Jonathan Cape Books, 1991).
2. Louise Morales-Brown, "Hyperthymesia: What Is It?," *Medical News Today*, September 21, 2023, https://www.medicalnewstoday.com/articles/hyperthymesia.
3. Chai M. Tyng, Hafeez U. Amin, Mohamed N.M. Saad, and Aamir S. Malik, "The Influences of Emotion on Learning and Memory," *Frontiers of Psychology* 8, no. 1454 (2017): 1–22. https://doi.org/10.3389/fpsyg.2017.01454.
4. G. Santos and V. Costa, "False Memory Syndrome: A Review and Emerging Issues, Following a Clinical Report," *European Psychiatry* 33, supplement (2016):S561. https://doi.org/10.1016/j.eurpsy.2016.01.2078.
5. A. Mattarella-Micke and S.L. Beilock, "Capacity Limitations of Memory and Learning," in *Encyclopedia of Memory and Learning*, ed. N.M. Seel (Boston, MA: Springer, 2012). https://doi.org/10.1007/978-1-4419-1428-6_603.
6. Saul McLeod, "Theories of Forgetting in Psychology," *Simply Psychology*, June 15, 2023, https://www.simplypsychology.org/forgetting.html.
7. Santos and Costa, "False Memory Syndrome."

Chapter 2: Bed Head
1. William Dement, "The Science of Dreams," *Newsweek* 53, November 30, 1959.
2. U.S. Centers for Disease Control and Prevention, "About Sleep," May 15, 2024, https://www.cdc.gov/sleep/about_us.html.
3. U.S. Centers for Disease Control and Prevention, "Impairments Due to Sleep Deprivation are Similar to Impairments Due to Alcohol Intoxication!" March 31, 2020, https://www.cdc.gov/niosh/work-hour-training-for-nurses/longhours/mod3/08.html.
4. "Extent and Health Consequences of Chronic Sleep Loss and Sleep Disorders," Chapter 3 in *Sleep Disorders and Sleep Deprivation: An Unmet Public Health Problem*, eds. H.R. Colten and B.M. Altevogt (National Academies Press, 2006). https://bit.ly/3O3xWLV.
5. Sources on sleep research disagree on whether there are three or four sleep stages prior to REM, but this is inconsequential for our purposes.
6. W.R. Klemm, "Why Does REM Sleep Occur? A Wake-Up Hypothesis," *Frontiers of Neuroscience* 5, no. 73 (2011). https://doi.org/10.3389/fnsys.2011.00073.

NOTES

7. David M. Eagleman and Don A. Vaughn, "The Defensive Activation Theory: REM Sleep as a Mechanism to Prevent Takeover of the Visual Cortex," *Frontiers in Neuroscience* 15:632853 (2021):1–10. https://doi.org/10.3389/fnins.2021.632853.

8. Brandon Peters, "How a Lack of REM Sleep Impacts Health and Learning," *Verywell Health*, December 6, 2024, https://www.verywellhealth.com/dream-deprivation-how-loss-of-rem-sleep-impacts-health-4159540.

9. University of Arizona Health Sciences, "An Epidemic of Dream Deprivation: Unrecognized Health Hazard of Sleep Loss," ScienceDaily, September 29, 2017, www.sciencedaily.com/releases/2017/09/170929093254.htm.

10. Bill Bryson, *The Body* (Anchor Books, 2019), 263–64.

11. "REM Sleep Behavior Disorder," *WebMD*, March 24, 2024, https://wb.md/3nxv5wF.

12. Mike Bribiglia, "Sleepwalking," *Just for Laughs* (YouTube Channel), May 21, 2019, https://www.youtube.com/watch?v=B02NsP33pRM.

13. Jade Wu, "Parasomnias: 6 Strange, Sometimes Creepy Sleep Phenomena," *Psychology Today*, January 28, 2021, https://www.psychologytoday.com/us/blog/the-savvy-psychologist/202101/parasomnias-6-strange-sometimes-creepy-sleep-phenomena.

14. Meir Kryger, *The Mystery of Sleep* (Yale University Press, 2017).

15. Most people do not dream while entering into Stage 1. I happen to be one who does, as I often awaken ten to fifteen minutes after dozing off, recall dreaming, and then shortly descend in earnest into a conventional sleep cycle.

16. Brandon Specktor, "Sleep Paralysis Is Linked to Stress (and Supernatural Beliefs)," *Live Science*, December 7, 2017, https://www.livescience.com/61123-sleep-paralysis-stress-supernatural.html.

17. Lucy Bryan and Abhinav Singh, "How to Lucid Dream: Expert Tips and Tricks," Sleep Foundation, September 20, 2024, https://www.sleepfoundation.org/dreams/lucid-dreams.

18. Naveed Saleh, "8 Ways to Trigger Lucid Dreaming," *Psychology Today*, November 16, 2016, https://bit.ly/3nY4lZA.

19. Karen R. Konkoly, "Real-Time Dialogue Between Experimenters and Dreamers During REM Sleep," Current Biology 31, no. 7 (2021):1417–27. https://doi.org/10.1016/j.cub.2021.01.026.

20. "Study Finds Real-Time Dialogue With a Dreaming Person Is Possible," *Medical Press*, February 18, 2021, https://medicalxpress.com/news/2021-02-real-time-dialogue-person.html.

21. Danielle Pacheco and Anis Rehman, "Nightmares in Children," Sleep Foundation, November 8, 2023, https://www.sleepfoundation.org/nightmares/nightmares-in-children.

22. Michael Schredl et al., "The Use of Dreams in Psychotherapy: A Survey of Psychotherapists in Private Practice," *Journal of Psychotherapy Practice and Research* 9, no. 2 (2000):81–87. https://pmc.ncbi.nlm.nih.gov/articles/PMC3330585/PMCID: PMC3330585.

23. G. William Domhoff, "Moving Dream Theory Beyond Freud and Jung," paper presented to the symposium "Beyond Freud and Jung?," Graduate Theological Union, Berkeley, CA, September 23, 2000. https://dreams.ucsc.edu/Library/domhoff_2000d.html.

NOTES

24. Matthias Forstmann and Pascal Burgmer, "Adults Are Intuitive Mind-Body Dualists," *Journal of Experimental Psychology General* 144, no. 1 (2015):222–35. doi:10.1037/xge0000045.

25. Poonam Sachdev, "Night Terrors," WebMD, December 6, 2023, https://www.webmd.com/sleep-disorders/night-terrors.

26. Zolar, *Zolar's Encyclopedia and Dictionary of Dreams* (Atria Books, 2010).

27. Mary Roach, *Spook: Science Tackles the Afterlife* (W.W. Norton and Co., 2005).

28. Suzana Herculano-Houzel, "The Human Brain in Numbers: A Linearly Scaled-Up Primate Brain," *Frontiers in Human Neuroscience* 3, no. 31 (2009):1–11. doi:10.3389/neuro.09.031.2009.

29. Ryan Whitwam, "New Transistor Mimics Human Synapse to Simulate Learning," Extreme Tech, November 6, 2013, https://www.extremetech.com/extreme/170411-new-transistor-mimics-human-synapse-to-simulate-learning.

30. Chris Woodford, "Neural Networks," Explain That Stuff, May 12, 2023, https://www.explainthatstuff.com/introduction-to-neural-networks.html.

31. For a number of fascinating stories on this and related topics, see *The Mind's Eye*, ed. Douglas R. Hofstadter (Bantam Books, 1981).

32. Forstmann and Burgmer, "Mind-Body Dualists."

33. For example, see Antonio Damasio, *Descartes' Error: Emotion, Reason, and the Human Brain* (Putnam & Sons, 1994).

34. Carlos A. Tinoca and João Ortiz, "Magnetic Stimulation of the Temporal Cortex: A Partial 'God Helmet' Replication Study," *Journal of Consciousness Exploration & Research* 5, no. 3 (2014):234–57.

35. M.A. Persinger et al., "The Electromagnetic Induction of Mystical and Altered States Within the Laboratory," *Journal of Consciousness Exploration & Research* 1, no. 7 (2010):808–30.

36. Andy Fell, "Psychedelic Drugs Change Structure of Neurons," *U.C. Davis News*, https://www.ucdavis.edu/news/psychedelic-drugs-change-structure-neurons.

37. Ruth Williams, "Neural Patterns of Consciousness Identified," *The Scientist*, February 6, 2019, https://www.the-scientist.com/neural-patterns-of-consciousness-identified-65433.

38. Kathleen A. Garrison et al., "Real-Time fMRI Links Subjective Experience With Brain Activity During Focused Attention," *Neuroimage* 81, no. 1 (2013):110–18. https://doi.org/10.1016/j.neuroimage.2013.05.030.

39. Jason Brownlee, "What Is Deep Learning?," Machine Learning Mastery, November 16, 2023, https://machinelearningmastery.com/what-is-deep-learning.

40. Alexx Kay, "Artificial Neural Networks," *Computer World*, February 12, 2001, https://www.computerworld.com/article/1361638/artificial-neural-networks.html.

41. This would be the perfect time to tell you the joke. But a funny thing happened between then and now: I forgot it. If I'd only had the precognitive foresight to know that one day I'd be writing about it in a book.

42. Rebecca Fuoco, "Spooky Nighttime Visitors: Paranormal or Parasomnia?," Project Sleep, no date, https://project-sleep.com/spooky-nighttime-visitors-paranormal-or-parasomnia.

43. This is similar to the experience of solving problems in dreams. See Hanan Parvez, "Problem-Solving in Dreams (Famous Examples)," *Psych Mechanics*, January 13, 2025, https://www.psychmechanics.com/problem-solving-in-dreams.

44. Certain drugs and mental practices can likely reveal some of what's going on in there. See Sam Harris, *Waking Up: A Guide to Spirituality Without Religion* (Simon & Schuster, 2014).

Chapter 3: Hanukkah Boys Wait Up for Santa

1. Quoted from an editorial in *The Sun* (New York), September 21, 1897.
2. Ray Bradbury, *Dandelion Wine* (Doubleday, 1957).
3. "Socialization," *The Free Dictionary*, https://encyclopedia.thefreedictionary.com/socialization.
4. "Trans-Cultural Diffusion," *The Free Dictionary*, https://encyclopedia.thefreedictionary.com/cultural+diffusion.
5. "History of Christmas Trees," History, March 2, 2025, https://www.history.com/topics/christmas/history-of-christmas-trees.
6. Quoted from the motion picture *A Christmas Story*, screenplay by Jean Shepherd, Bob Clark, and Leigh Brown, MGM Productions.
7. Bible, King James Version, 1 Corinthians 13:11.

Chapter 4: You Sneeze, You Die

1. Emo Philips, "The Best God Joke Ever—and It's Mine!" *The Guardian*, September 29, 2005, https://www.theguardian.com/stage/2005/sep/29/comedy.religion.
2. "Mother and Son Strangled, Father Critically Wounded," Wicked Local, July 19, 2015; originally from *Framingham News*, September 20, 1963, https://www.wickedlocal.com/story/bulletin-tab/2015/07/19/revisiting-double-murder-in-framingham/33846699007.
3. "Neighbors Recall Framingham Murderer," Wicked Local, August 17, 2015, https://www.wickedlocal.com/story/bulletin-tab/2015/08/16/neighbors-recall-framingham-murderer/33670647007.
4. "*Abington School District v. Schempp*, 374 U.S. 203 (1963)," Justia, https://supreme.justia.com/cases/federal/us/374/203/.
5. Jaweed Kaleem, "School Prayer 50 Years Later: What Do Americans Believe?," *Huffington Post*, December 6, 2017, https://www.huffpost.com/entry/school-prayer_n_3461479.
6. "Does Your Heart Stop When You Sneeze?," Library of Congress, May 6, 2024, https://www.loc.gov/everyday-mysteries/biology-and-human-anatomy/item/does-your-heart-stop-when-you-sneeze/.
7. Carl Sagan, *Broca's Brain: Reflections on the Romance of Science* (Ballantine Books, 2014), 335.
8. Richard Dawkins, *The God Delusion* (Houghton Mifflin, 2006), chapter 3. Here Dawkins summarizes the counterarguments more accessibly than the philosophers.
9. True, some gods' versions of morality are morally repugnant to the followers of other gods. A god's morality can even be repugnant to its own followers. In addition to being extremely thin-skinned and narcissistic, the gods of even the most popular religions

have been known to condone slavery, perversion, lying, murder, cannibalism, incest, and baby-killing. Most religions are quite clear about devaluing the lives of people who committed the sin of being born into the wrong religion.

Chapter 5: U.F. ... Ohhh?

1. Neil deGrasse Tyson, *Death by Black Hole: And Other Cosmic Quandaries* (W.W. Norton & Co., 2007).
2. Robert Sheaffer, *UFO Sightings: The Evidence* (Prometheus Books, 1998).
3. Quite recently the term *unidentified aerial phenomenon* (UAP) has gained traction among experts and those who fancy themselves as such. It was designed to take away some of the extraterrestrial connotations of UFO. I doubt that changing the label will accomplish that. Regardless, I'll continue to use UFO because it's still the more broadly understood acronym at the time I'm writing this.
4. Mick West, direct message to author, June 6, 2022. See Mick's metabunk.org site for some of his in-depth analyses of UFO evidence and other anomalous claims.
5. This illustrates *Occam's Razor*, which argues for provisionally accepting the simplest explanation over alternatives that invoke assumptions about unknown or unnecessarily complex causes.
6. James V. Hart et al., *Contact* (screenplay), Warner Brothers, 1997. Based on Carl Sagan's book *Contact* (Pocket Books, 1986).
7. "Michael Shermer," Wikipedia, accessed March 20, 2025, https://en.wikipedia.org/wiki/Michael_Shermer.
8. Barry Markovsky, "UFOs," in *Skeptic Encyclopedia of Pseudoscience*, 1 (2002):260–71 (ABC-CLIO, 2002).
9. Sheaffer, *UFO Sightings*.
10. Philip J. Klass, *UFOs: The Public Deceived* (Prometheus Books, 1983).
11. Mick West, "I Study UFOs—and I Don't Believe the Alien Hype. Here's Why," *The Guardian*, June 11, 2021, https://www.theguardian.com/commentisfree/2021/jun/11/i-study-ufos-and-i-dont-believe-the-alien-hype-heres-why. Mick didn't actually start publishing skeptical analyses until a few years after my encyclopedia article was published. But today his name absolutely deserves to be mentioned alongside the likes of Klass and Sheaffer.
12. "Do Americans Believe in UFOs?," *Gallup News*, August 20, 2021, https://news.gallup.com/poll/350096/americans-believe-ufos.aspx.
13. Eir Nolsoe, "Half of Britons Think Aliens Exist—and 7% Claim to Have Seen an UFO," *YouGov U.K.*, June 25, 2021, https://yougov.co.uk/politics/articles/36619-half-britons-think-aliens-exist-and-7-claim-have-s.
14. Nicole Mortillaro, "Four in Five Canadians Believe in Aliens: Angus Reid Poll," *Global News*, August 24, 2016, https://globalnews.ca/news/2900372/four-in-five-canadians-believe-in-aliens-angus-reid-poll/.
15. The report has since been finalized and updated several times, with "no evidence of extraterrestrial beings, activity, or technology" ever being found. The most recent report at this writing: "Fiscal Year 2024 Consolidated Annual Report on Unidentified Anomalous Phenomena," Department of Defense All-Domain Anomaly Resolution Office,

https://www.dni.gov/files/ODNI/documents/assessments/DOD-AARO-Consolidated-Annual-Report-on-UAP-Nov2024.pdf.

16. Sharon A. Hill, *Scientifical Americans* (McFarland & Company, 2017).

17. MUFON—Mutual UFO Network, https://mufon.com/.

18. Robert Sheaffer, *Bad UFOs: Critical Thinking About UFO Claims* (CreateSpace Independent Publishing, 2016).

Chapter 6: All Rise

1. J.M. Barrie, *The Complete Adventures of Peter Pan* (Benediction Classics, 2013).

2. Susane Colasanti, *When It Happens* (Viking Juvenile, 2006). Quoted at https://www.goodreads.com/quotes/306318-standing-in-the-line-at-the-food-court-i-try.

3. Lindsey Leavitt, *Going Vintage* (Bloomsbury, 2013). Quoted at https://www.goodreads.com/author/quotes/2945688.Lindsey_Leavitt.

4. Dustin Albert et al., "The Teenage Brain: Peer Influences on Adolescent Decision Making," *Current Directions in Psychological Science* 22, no. 2 (2013):114–20, https://doi.org/10.1177%2F0963721412471347.

5. John F. Kihlstrom, "Is Hypnosis an Altered State of Consciousness *Or What?*" *Contemporary Hypnosis* 22, no. 1 (2005):34–38.

6. Kirsten Weir, "Uncovering the New Science of Clinical Hypnosis," *Monitor on Psychology* (American Psychological Association) 55, no. 3 (2024):27. https://www.apa.org/monitor/2024/04/science-of-hypnosis.

7. Mary Roach, *Spook* (W.F. Norton, 2005).

8. Gabriel Andrade, "Is Past Life Regression Therapy Ethical?," *Journal of Medical Ethics and History of Medicine* Dec. 2, no. 10 (2017):11. PMID: 29416831; PMCID: PMC5797677.

9. Harriet Hall, "Hypnosis Revisited," *Skeptical Inquirer* 45, no. 2 (2021). https://skepticalinquirer.org/2021/03/hypnosis-revisited/.

10. Rory O'Neill Schmitt and Rosary Hartel O'Neill, *New Orleans Voodoo: A Cultural History* (The History Press, 2019).

11. David W. Moore, "Three in Four Americans Believe in Paranormal," *Gallup News*, June 16, 2005, https://news.gallup.com/poll/16915/three-four-americans-believe-paranormal.aspx.

12. Tracy V. Wilson, "How Voodoo (Vodou) Works," How Stuff Works, May 17, 2022, https://people.howstuffworks.com/voodoo.htm#pt4.

13. Brian Salter, "Can Voodoo Doll Spells Hurt Someone," Quora, April 29, 2017, https://www.quora.com/Can-Voodoo-doll-spells-hurt-someone/answer/Brian-Salter-1.

14. "Ideomotor Effect," Skeptic's Dictionary, September 12, 2014, http://www.skepdic.com/ideomotor.html.

15. "Do You Believe?," *National Geographic Brain Games* (YouTube channel), February 5, 2015, https://www.youtube.com/watch?v=PRo8TytvIDw.

16. Derren Brown, "Séance—Full Episode," *Derren Brown* (YouTube channel), March 7, 2020, https://www.youtube.com/watch?v=KFj3L28zoqs.

17. Haily Reese, "*WARNING* Light as a Feather, Stiff as a Board," *Haily Reese* (YouTube channel), July 24, 2020, https://www.youtube.com/watch?v=6I3nMHV7a34&t=525s.

18. "Levitating Man Trick Revealed," *The Q* (YouTube channel), https://www.youtube.com/watch?v=87thN8tVvgI.

19. "David Copperfield: Flying," *Calendata* (YouTube channel), June 6, 2013, https://www.youtube.com/watch?v=OoRuoteg5sE.

20. "The Magic of David Copperfield VI: Floating Over the Grand Canyon—1984," *MagicWeek* (YouTube channel), https://www.youtube.com/watch?v=aynIKJE2QAs&t=344s.

21. "US patent #5354238A Levitation Apparatus," Espacenet Patent Search, accessed March 25, 2025, https://worldwide.espacenet.com/patent/search/family/025327463/publication/US5354238A?q=pn%3DUS5354238.

22. It's evident in the Grand Canyon video that several different segments used more conventional lifts and that camera tricks were added in the editing.

23. Gemma Perry et al., "Rhythmic Chanting and Mystical States Across Traditions," *Brain Science* 11, no. 1 (2021):1–17, https://doi.org/10.3390/brainsci11010101.

Chapter 7: Onward Christian Scientists

1. Richard Nordquist, "*The Lowest Animal* by Mark Twain," Thought Co., April 27, 2019. https://www.thoughtco.com/the-lowest-animal-by-mark-twain-1690158. The quote originally appeared in an essay written in 1896, published in different forms under different titles.

2. "The Common Cold," CBS TV, December 29, 1965. See listing at IMDB, https://www.imdb.com/title/tt0522605/.

3. Stephen Barrett, "The Origin and Current Status of Christian Science," Quackwatch, March 16, 2016, https://quackwatch.org/related/cs2/.

4. Rodney Stark, 1998. "The Rise and Fall of Christian Science," *Journal of Contemporary Religion* 13, no. 2 (1998):189–214. https://doi.org/10.1080/13537909808580830.

5. Andrew Skolnick, "Christian Scientists Claim Healing Efficacy Equal If Not Superior to That of Medicine," *Journal of the American Medical Association* 264, no. 11 (1990):1379–81. doi:10.1001/jama.1990.03450110015003.

6. Skolnick, Christian Scientists, 1379.

7. "What Is Christian Science Treatment?," *The Christian Science Monitor*, August 2, 1990, https://www.csmonitor.com/1990/0802/mrc291.html.

8. David Margolick, "In Child Deaths, a Test for Christian Science," *The New York Times*, August 6, 1990, https://www.nytimes.com/1990/08/06/us/in-child-deaths-a-test-for-christian-science.html.

9. Rita Swan, "Victims of Religion-Based Medical Neglect," Children's Health Care, accessed March 26, 2025, https://childrenshealthcare.org/victims/.

10. Caroline Fraser, "Dying the Christian Science Way: The Horror of My Father's Last Days," *The Guardian*, August 6, 2019, https://www.theguardian.com/world/2019/aug/06/christian-science-church-medicine-death-horror-of-my-fathers-last-days.

11. William Franklin Simpson, "Comparative Longevity in a College Cohort of Christian Scientists," *Journal of the American Medical Association* 262, no. 12 (1989):1657–58, Doi:10.1001/jama.1989.03430120111031.

NOTES

12. Elise Wolff, "The Christian Science Church in the Twenty-first Century," *Journal of Contemporary Religion* 35, no. 3 (2020):565–73, http://dx.doi.org/10.1080/13537903.2020.1811538.

13. Stephen Barrett, "Christian Science Statistics: Practitioners, Teachers, and Churches in the United States," Quackwatch, March 13, 2016, https://quackwatch.org/related/cs/.

14. Donald Prothero and Timothy D. Callahan, *UFOs, Chemtrails, and Aliens: What Science Says* (Indiana University Press, 2017), 165.

15. Matthew Thorpe, "How Meditation Benefits Your Mind and Body," *Healthline*, August 15, 2024, https://www.healthline.com/nutrition/12-benefits-of-meditation.

16. Christopher G. Ellison et al., "Prayer, Attachment to God, and Symptoms of Anxiety-Related Disorders among U.S. Adults," *Sociology of Religion* 75, no. 2 (2014):208–33, https://doi.org/10.1093/socrel/srt079. However, the authors find evidence suggesting that the mental health benefits to the individual are contingent on the praying person's attachment relationship with their god, e.g., anxious versus secure.

17. Robin Nusslock et al., "Higher Peripheral Inflammatory Signaling Associated with Lower Resting-State Functional Brain Connectivity in Emotion Regulation and Central Executive Networks," *Biological Psychiatry* 86, no. 2 (2019):153–62, doi:10.1016/j.biopsych.2019.03.968.

18. "2023–24 U.S. Religious Landscape Study Interactive Database," Pew Research Center, 2025, doi:10.58094/3zs9-jc14.

19. David A. Hodge, "Systematic Review of the Empirical Literature on Intercessory Prayer." *Research on Social Work Practice* 17, no. 2 (2007):174–87, https://doi.org/10.1177/1049731506296170.

20. The seven studies that showed significant prayer effects all had potential shortcomings and had not been replicated by independent researchers. That makes them interesting but still inconclusive.

21. H. Benson et al., "Study of the Therapeutic Effects of Intercessory Prayer (STEP) in Cardiac Bypass Patients: A Multicenter Randomized Trial of Uncertainty and Certainty of Receiving Intercessory Prayer," *American Heart Journal* 151, no. 4 (2006):934–42, https://doi.org/10.1016/j.ahj.2005.05.028.

22. Morris Zelditch, Jr., "Processes of Legitimation: Recent Developments and New Directions," *Social Psychology Quarterly* 64, no. 1 (2001):4–17, https://doi.org/10.2307/3090147.

23. Troy DuFrene, "Two Boats and a Helicopter: Thoughts on Stress Management," *Psychology Today*, May 4, 2009, https://www.psychologytoday.com/us/blog/fumbling-for-change/200905/two-boats-and-a-helicopter-thoughts-on-stress-management.

CHAPTER 8: HE'S PULLING HER LEG

1. Mark Vernon, "William James, Part 4: The Psychology of Conversion," *The Guardian*, November 8, 2010, https://www.theguardian.com/commentisfree/belief/2010/nov/08/psychology-religious-conversion.

2. Barry Markovsky and Jake Frederick, "Mysticism," in *Emerging Trends in the Social and Behavioral Sciences*, ed. Robert A. Scott (Wiley, 2015), https://doi.org/10.1002/9781118900772.etrds0231.

NOTES

3. Sarah M. Fischer, "What Is Love Bombing, and What Does It Look Like?," *Nebraska Medicine*, August 18, 2013, https://www.nebraskamed.com/health/conditions-and-services/behavioral-health/what-is-love-bombing-and-what-does-it-look-like.

4. The Church of Scientology is especially renowned for using intimidation and litigation against their ex-members. See Richard Behar, "The Thriving Cult of Greed and Power," *Time Magazine*, May 6 (1991): 50, https://www.cs.cmu.edu/~dst/Fishman/time-behar.html.

5. Michael T. Hannan and John Freeman, *Organizational Ecology* (Harvard University Press, 1989), 3–27.

6. "Social Impact Theory," in *Encyclopedia of Social Psychology*, ed. Roy F. Baumeister and Kathleen D. Vohs (Sage Publications, 2007), doi:10.4135/9781412956253.n534.

7. My colleague and I did as much in a lab setting. See Barry Markovsky and Shane Thye, "Social Influences on Paranormal Beliefs," *Sociological Perspectives* 44, no. 1 (2001):21–44, https://doi.org/10.1525/sop.2001.44.1.21. We also published a less technical version: "Social Influence and the Power of the Pyramid," *Skeptic* 9, no. 3 (2002):36–41.

8. Something like this actually happened to Delta Air Lines Flight 723 at Boston's Logan Airport on July 31, 1973. I don't know if the lone survivor attributed his luck to God because he was severely burned and died some months later from complications having lived out his grace time in considerable agony. But many in the public believed it to have been a miraculous event. Eerily, before takeoff, another passenger on the same flight asked to deboard the plane after it was already on the taxiway. The Captain accommodated his request, and so he was spared from the tragedy. But it wasn't because the passenger had a premonition. It was because he realized that he was going to miss a business meeting later that day, making his flight unnecessary. I suppose one could construe this as another sort of miracle, but it feels like a bit of a stretch. See "Delta Flight 723 Crashes in Fog at Boston Logan," History, June 27, 2023, https://www.history.com/this-day-in-history/delta-flight-723-crashes-in-fog-at-boston-logan.

9. This is a logical fallacy, aptly named the *argument from ignorance*.

10. James Randi, "Be Healed in the Name of God," *Free Inquiry* 6, no. 2 (1986):8–19.

11. "An Honest Liar," Wikipedia, accessed March 26, 2025, https://en.wikipedia.org/wiki/An_Honest_Liar.

12. James Randi, *The Faith Healers* (Prometheus Books, 1987). Randi investigated a number of faith healers and their various deceptions. He looked into both the prominent and the obscure, never finding evidence of any actual healing, always finding evidence of misdirection and deception. As a conjurer himself, Randi knew what to look for.

13. "Motivated Cognition," Psychology, undated, https://psychology.iresearchnet.com/social-psychology/social-cognition/motivated-cognition/.

CHAPTER 9: ASTRONOLOGY

1. Arthur C. Clarke, *3001: The Final Odyssey* (Harper Collins, 1989).

2. You may have noticed that sometimes I connect myself with psychology, other times sociology. Funny story about that. Officially, I was a psychology major at UMass but equally interested in sociology. The university recognized neither double-majors nor minors. This bothered me because, having foregone getting any sleep in the fall of my senior year, I had satisfied all the graduation requirements in both fields by the end of

NOTES

the semester. I could graduate! I'd also completed honors projects in both departments. I received a higher honors designation in my non-major of sociology than I did in psychology, but the higher honor (*magna cum laude*) would not appear on my transcript. I decided to try to get the Bureaucracy to make it happen. It took most of a year. Fortunately, my advisors were all very supportive, including Drs. Anthony Harris in Sociology, Seymour Berger in Psychology, and George Armelagos, who was the Honors Program Director. Eventually, we got our way, and I became the first student in UMass history to change majors *after* graduating. That honor plus a few bucks will get me a cup of coffee.

3. Cary Funk and Sara Kehaulani Goo, "A Look at What the Public Knows and Does Not Know About Science," Pew Research Center, September 10, 2015, https://www.pewresearch.org/science/2015/09/10/what-the-public-knows-and-does-not-know-about-science/.

4. Ida Andersson et al., "Even the Stars Think That I Am Superior: Personality, Intelligence and Belief in Astrology," *Personality and Individual Differences* 187 (2022):1–3, http://dx.doi.org/10.1016/j.paid.2021.111389.

5. For a great overview, see Geoffrey Deans articles, "Does Astrology Need to Be True? Part 1: A Look at the Real Thing," *Skeptical Inquirer* 11, Winter (1986):166–84; and "Does Astrology Need to Be True? Part 2: The Answer Is No," *Skeptical Inquirer* 11, Spring (1987):257–73. And in case you're wondering, Indian and Chinese Astrology fare no better in tests. See Jayant Narlikar, "An Indian Test of Indian Astrology," *Skeptical Inquirer* 37, no. 2 (2013):45–49.

6. For an extensive compilation of astrology research published as of 2016, see https://www.astrology-and-science.com. For an excellent older review: Terence Hines, *Pseudoscience and the Paranormal* (Prometheus Books, 1988), chapter 6.

7. J. Mayo et al., "An Empirical Study of the Relation Between Astrological Factors and Personality," *Journal of Social Psychology* 105, no. 2 (1978):229–36, https://psycnet.apa.org/doi/10.1080/00224545.1978.9924119.

8. Derek Schaedig, "Self-Fulfilling Prophecy in Psychology: Definition & Examples," Simply Psychology, February 13, 2024, https://www.simplypsychology.org/self-fulfilling-prophecy.html.

9. Hines, *Pseudoscience and the Paranormal*.

10. For a summary of Gauquelin's study, see Jan Willem Nienhuys, "The Mars Effect in Retrospect," *Skeptical Inquirer*, Nov./Dec. (1997):24–29.

11. Ibid., 26–28.

12. Shawn Carlson, "A Double-Blind Test of Astrology," *Nature* 318, no. 5 (1985): 419–25.

13. Like the Carlson study, a strict double-blind was maintained. This means that Celeste's mother had no knowledge (beyond birth information) of the subject of her natal reading, nor did Celeste, the one person from the study with whom she had any contact.

14. Daisie Radner and Michael Radner, *Science and Unreason* (Wadsworth Publishing Co., 1982).

15. Devrupa Rakshit, "Why Do People Still Believe in Astrology?," The Swaddle, June 24, 2020, https://theswaddle.com/why-do-people-still-believe-in-astrology/.

NOTES

16. My demo was loosely based on James Randi's classroom application, which can be viewed here: https://www.youtube.com/watch?v=3Dp2Zqk8vHw. The reverse-worded version that I used for half of my students was unique to my version.

17. I don't recall where I read it, but I believe it's true! See "The 6 Most Gullible Zodiac Signs Who Are Way Too Trusting," Your Tango, April 10, 2018, https://www.yourtango.com/2018311923/horoscope-six-gullible-zodiac-signs-who-are-too-trusting-according-astrology.

Chapter 10: You Made the Earth Move

1. "Earthquake Safety Tips," *The Onion*, October 6, 1999, https://theonion.com/earthquake-safety-tips-1819565330/.

2. Scientific method means very carefully checking your claims and explanations. There are two kinds of checks: logical integrity and empirical accuracy. Logical integrity means that your explanation cannot be self-contradictory, vague, or leap to unwarranted conclusions. Empirical accuracy means that your evidence is sufficiently fine-grained and precise to support your claim and to rule out alternative accounts. Scientific method is optional in science *only* if you're okay with contradicting yourself, overgeneralizing your claim, or stating something as fact without sufficient evidence.

3. Barry Markovsky, "Graduate Training in Sociological Theory and Theory Construction." *Sociological Perspectives*, 51, no. 2 (2008):423–47.

4. Dan Pilat and Sekoul Krastev, "Why Do We Exaggerate Some Details of a Story, But Minimize Others?," The Decision Lab (undated), https://thedecisionlab.com/biases/leveling-and-sharpening.

5. The calculation goes like this. On a given day:

$p_e = 0.0328$	prob. of a feelable earthquake = 12/365
$p_d = 0.0657$	prob. of an earthquake dream = 24/365
$p_{d\&e} = 0.00216$	prob. of both of the above on a given day = $p_e \times p_d$
$1 - p_{d\&e} = 0.9979$	prob. of neither dream nor earthquake on a given day
$(1 - p_{\&e})^{365} = 0.4539$	prob. neither occurs in 365 days
$1 - (1 - p_{d\&e})^{365} = 0.5461$	prob. both occur together at least once in 365 days

6. Astute readers may be reminded of the "birthday paradox": In a random group of twenty-three people, there's over a 50 percent chance that two will have the same birthday. The set-up for the calculations is a little different, but the result is surprising for similar psychological and mathematical reasons.

7. Ursula K. Le Guin, *The Lathe of Heaven* (Avon Books, 1971).

8. Michael Shermer, *The Believing Brain* (Times Books, 2011), chapters 4–5.

9. See also Bruce M. Hood's *Supersense: Why We Believe in the Unbelievable* (HarperOne, 2009).

10. "The Loma Prieta Legacy," Stanford 125, undated, https://125.stanford.edu/the-loma-prieta-legacy/.

Chapter 11: Dowsing the Dowser

1. Charles Fort, *The Complete Books of Charles Fort* (Courier Corporation, 1941).

2. "The American Society of Dowsers," https://dowsers.org/.

NOTES

3. Fair Warning: This note ventures into some statistical weeds.

To feel safe in my wager with Vern, I'd calculated P_x in advance—the probability of getting exactly x successes across some number of trials when the probably of success in *each* trial is p. The calculation is a standard one in probability theory but not so simple.

In our case, there were four ways for Vern to succeed: being correct seven, eight, nine, or ten times out of ten tries. Any number of successes less than seven would have been considered failing. Calculating P_x then requires solving the problem for each of those four outcomes—exactly seven successes, exactly eight successes, and so on. Then we can add those four probabilities together to find out the likelihood of success expected by mere guessing.

First, we need to know how many combinations of ten tries can produce each of those four outcomes. Just as when rolling two dice there are multiple ways to get a five (i.e., one and four, two and three, three and two, and four and one) but only one way to get a twelve (six and six), which makes a five a lot more likely. Vern had multiple ways to get seven, eight, or nine correct. There's only one way for him to get ten successes in ten tries: Every trial must be a success. He could get nine correct ten different ways. That is, he could make his one wrong guess on the first trial, or on the second, third, fourth, and so on.

To calculate the number of ways to get seven or eight correct, we use the formula for *combinations*:

$$C_{n,x} = n! / x!(n-x)!$$

This is read "The number of combinations of n objects chosen x at a time is...." The "!" means "factorial," which is calculated this way: $q! = q \times (q-1) \times (q-2) \times \ldots \times 1$. So, for example, $4! = 4 \times 3 \times 2 \times 1 = 24$. Applying this formula in our problem, there are 45 ways to get 8 correct in 10 tries, and 120 ways to get seven correct in 10 tries.

The *binomial probability* formula takes into account the probability of success versus failure on each try. One cup had pennies, two didn't, so the probability of guessing right on any of the ten trials is one in three, or $p = 0.3333$. We make $q = 0.6666$ (which is just $1 - p$) the probability of failing on each trial. Applying all this math to our four outcomes of interest (x = seven, eight, nine, or ten correct) looks like this:

$$P_x = C_{n,x} \, p^x Q^{n-x}$$

This gives us

$$P_7 = 0.01624, \; P_8 = 0.00304, \; P_9 = 0.00034, \; P_{10} = 0.00002$$

Adding them all up,

$$\text{Total: } 0.01964 \approx 2\%$$

4. I used a random number table ahead of time to generate the ten sequences and brought them with me on a slip of paper. This was because humans are not good randomizers. See Małgorzata Figurska et al., "Humans Cannot Consciously Generate Random Number Sequences: Polemic Study," *Medical Hypotheses* 70, no. 1 (2008):182–85.

5. R.A. Foulkes, "Dowsing Experiments," *Nature* 229 (1971):163–68, https://doi.org/10.1038/229163a0.

6. J.T. Enright, "Testing Dowsing: Failure of the Munich Experiments," *Skeptical Inquirer* 23, no. 1 (1999):39–46.

NOTES

7. James Randi, "A Controlled Test of Dowsing Abilities," *Skeptical Inquirer* 4, no. 1 (1979):16–20.

8. "Water Table," Wikipedia, accessed March 26, 2025, https://en.wikipedia.org/wiki/Water_table.

9. "Ideomotor Effect," The Skeptic's Dictionary, September 12, 2014, http://www.skepdic.com/ideomotor.html.

10. It's also what moves the planchet on a Ouija board. See "Dowsing (a.k.a. Water Witching, Radiesthesia)," The Skeptic's Dictionary, October 29, 2015, http://www.skepdic.com/dowsing.html.

11. Camila Domonoske, "U.K. Water Companies Sometimes Use Dowsing Rods to Find Pipes," National Public Radio, November 21, 2017, https://www.npr.org/sections/thetwo-way/2017/11/21/565746002/u-k-water-companies-sometimes-use-dowsing-rods-to-find-pipes.

12. "Drug, Weapon Tracker Yanked Off the Market," *Orlando Sentinel*, July 31, 2021, https://www.orlandosentinel.com/1996/01/25/drug-weapon-tracker-yanked-off-the-market/.

13. Phil Plait, "Maker of Useless Dowsing Rod for Bombs Convicted for Fraud," *Slate*, April 29, 2013, https://slate.com/technology/2013/04/dowsing-for-bombs-maker-of-useless-bomb-detectors-convicted-of-fraud.html.

14. Johan Spanner, "Iraq Swears by Bomb Detector U.S. Sees as Useless," *The New York Times*, November 3, 2009, https://www.nytimes.com/2009/11/04/world/middleeast/04sensors.html.

CHAPTER 12: THE YOGI HAS NO ROBES

1. Sachin Garg, *Come on Inner Peace! I Don't Have All Day!* (Grapevine Publishers, 2013).

2. "What Is the *Transcendental Meditation* Technique?," Maharishi Foundation International, https://www.tm.org/en-us/what-is-tm.

3. "Beatles Guru Maharishi Mahesh Yogi Dies," *The Sydney Morning Herald*, February 7, 2008, https://www.smh.com.au/world/beatles-guru-maharishi-mahesh-yogi-dies-20080206-1qno.html. For the rest of his life, the Maharishi was perhaps best known as guru to the Beatles. There are conflicting accounts on the Internet as to whether or not the Beatles became disillusioned with the Maharishi.

4. Maharishi Mahesh Yogi, *Maharishi's Programs to Create Heaven on Earth* (Age of Enlightenment Books, circa 1992).

5. Jennifer 8. Lee, "Peace, and Kucinich, Gets a Chance," *The New York Times*, January 18, 2004, https://www.nytimes.com/2004/01/18/style/peace-and-kucinich-gets-a-chance.html. In fact, the domes aren't gilded, or even metallic. They're coated in gold-colored spray foam.

6. At some point, MIU increased dome security and surrounded them with high fences. Peeking inside on a random Saturday would not be so easy today.

7. Robert A. Rabinoff, Michael C. Dillbeck, and Robert Deissler, "Effect of Coherent Collective Consciousness on the Weather," paper presented at the 1982 Annual Meeting of the Iowa Academy of Science, Fort Dodge, IA.

8. This is my redrawn, simplified version of a graph appearing on p. 795 in David W. Orme-Johnson et al., "International Peace Project in the Middle East: The Effects of the Maharishi Technology of the Unified Field," *Journal of Conflict Resolution* 32, no. 4 (1988):776–812, https://www.jstor.org/stable/174032. It was based on my hand measurements of their published graph because the authors would not share their actual data with me. More on that later.

9. I thought Bill Bryson captured this brilliantly in *A Short History of Nearly Everything* (Crown, 2004).

10. To be clear about the analysis to follow, when I cite evidence, I'm making factual claims. When I characterize the evidence, I'm giving my honest, informed opinion. And when I describe my own or others' experiences and impressions, they are only that.

11. "Higher States of Consciousness 16-Lesson Course," Maharishi International University, accessed March 27, 2025, https://www.miu.edu/higher-states-ce-course.

12. Researchers at Johns Hopkins University reviewed the published literature on several major types of meditation. They looked at health outcomes such as stress, drug use, and depression. Out of hundreds of published TM-based studies, *only eight* satisfied minimal standards for clinical research. Among those eight, TM showed no benefits beyond other forms of meditation. Also, TM fell short of mindfulness meditation for several health outcomes. Note that learning TM incurs a hefty fee, whereas other methods can be learned at no cost. See Madhav Goyal et al., "Meditation Programs for Psychological Stress and Well-Being: A Systematic Review and Meta-analysis," *Journal of the American Medical Association Internal Medicine* 174, no. 3 (2014):357–68, doi:10.1001/jamainternmed.2013.13018

13. If what emanates from the meditation group truly affects the unified field of matter and energy, then surely it could move the needle on a purpose-built Consciousness Field Detector. Instead, all of the dozens of Maharishi Effect studies exclusively use indirect measures, opening the door to other unmeasured causes and alternative explanations.

14. Franklin D. Trumpy, "An Investigation of the Reported Effect of Transcendental Meditation on the Weather," *Skeptical Inquirer* 8, Winter (1983):143–48.

15. Robert A. Rabinoff et al., "Effect of Coherent Collective Consciousness on the Weather," *Scientific Research on Maharishi's Transcendental Meditation and the TM-Sidhi Programme* 4, no. 324 (1989):2564–65.

16. "Transcendental Meditation Research Collected Papers," Maharishi International University, https://research.miu.edu/collected-papers/.

17. Nicole Napoli, "A Nap a Day Keeps High Blood Pressure at Bay," American College of Cardiology, March 7, 2019, https://www.acc.org/about-acc/press-releases/2019/03/07/08/56/a-nap-a-day-keeps-high-blood-pressure-at-bay.

18. Soo Liang Ooi et al., "Transcendental Meditation for Lowering Blood Pressure: An Overview of Systematic Reviews and Meta-Analyses," *Complimentary Therapies in Medicine* 34, October (2017):26–34, https://doi.org/10.1016/j.ctim.2017.07.008.

19. A group of ten thousand Siddhi meditators convened in India at the start of 2024 explicitly to bring about "immediate world peace." Meanwhile, wars continued unabated in Palestine, Ukraine, Sudan, Yemen, and elsewhere around the world.

NOTES

20. R. Nuzzo, "How Scientists Fool Themselves—and How They Can Stop," *Nature* 526, October (2015):82–185, https://doi.org/10.1038/526182a. Also see "Case Summaries" listed by the Office of Research Integrity, https://ori.hhs.gov/content/case_summary.

21. Eight measures were combined into a single quality-of-life measure. These were chosen over other possibilities without any clear rationale. We have to consider the possibility they were cherry-picked to produce the desired results—an unethical practice disparagingly referred to as "drawing the bullseye around the arrow." At the very least, the TM researchers must let others check their results and test alternative measures, but they won't. More on this below.

22. See Evan Fales and Barry Markovsky, "Evaluating Heterodox Theories," *Social Forces* 7(2):511–25, https://doi.org/10.2307/2580722.

23. This is usually attributed to Nobel Prize–winning economist Ronald H. Coase.

24. Mordecai Kaffman, "The Use of Transcendental Meditation to Promote Social Progress in Israel," *Cultic Studies Journal* 3, no. 1 (1986):97–102. See also, Aryeh Siegel, *Transcendental Deception* (Janreg Press, 2018).

25. One of my claims about the research was that they made "selective use of evidence." Is this "derogatory," or merely critical? Regardless, it's not false. I explained this in writing to lead investigator David Orme-Johnson with one of many possible examples:

> In the *Des Moines Register* . . . Professor Charles Alexander gave MIU credit for the Berlin Wall coming down, the fall of the Soviet Union, and the fall of Apartheid. He was identified as an MIU scientist and researcher, and stated that he was "really confident that we are the cause" of these events. During the same historical period, many, many unfortunate and tragic events also occurred, and Dr. Alexander neglected to mention any of them. He was therefore making selective use of evidence.

In an article at Truth About TM (https://www.truthabouttm.org/SocietalEffects/Critics-Rebuttals/index.cfm#Israel_Data), Orme-Johnson claimed both that he *did* share the Jerusalem data with me—as graphs—and also that he *didn't* share the data—as numerical spreadsheets—knowing that I'd need the numbers to check his results. For me, such rhetorical game-playing makes the authors' research even less plausible.

26. Philip A. Schrodt, "A Methodological Critique of a Test of the Effects of the Maharishi Technology of the Unified Field," *Journal of Conflict Resolution* 34, no. 4 (1990):745–55, https://doi.org/10.1177/0022002790034004008.

27. They could have even used the same quality-of-life measures, except they should reveal them prior to the research and state exactly how they'd be measured—not a standard practice for them.

28. Siegel, *Transcendental Deception.*

29. Some of this work is discussed in the last section of "A Methodological Review of Maharishi Effect Research," Maharishi International University, accessed March 27, 2025, https://research.miu.edu/maharishi-effect/a-methodological-review-of-maharishi-effect-research.

30. Most famously, this so-called *epistemic corruption* has impacted medical science via the pharmaceutical industry. See Sergio Sismondo, "Epistemic Corruption, the

Pharmaceutical Industry, and the Body of Medical Science," *Frontiers in Research Metrics and Analytics* 6, March (2021), https://doi.org/10.3389/frma.2021.614013.

31. Herbert Benson and Miriam Z. Klipper, *The Relaxation Response* (William Morrow Paperbacks, 2000).

32. "Revolutionising Peace: 10,000 Minds in Sync Paves the Way for Immediate World Peace," Daiji World, https://www.daijiworld.com/news/newsDisplay?newsID=1157447.

33. Pepijn AL, "The Value of Communities and Their Consent: A Communitarian Justification of Community Consent in Medical Research," *Bioethics* 35, October (2020):255–61, https://doi.org/10.1111/bioe.12820.

34. Siegel, *Transcendental Deception,* Chapter 12.

35. "Fairfield, IA Crime Rates," Neighborhood Scout, accessed June 27, 2024, https://www.neighborhoodscout.com/ia/fairfield/crime. Also, see Siegel, *Transcendental Deception*, chapter 7 https://www.tmdeception.com/bogusresearch.

36. Dr. Yannick Meurice, personal communication to author, October 17, 1996.

37. Robert L. Park, "Voodoo Science and the Belief Gene," *Skeptical Inquirer*, Sept./Oct. (2000):24–29.

38. Michael Shermer, *Why People Believe Weird Things* (Holt eBook, 2002): Kindle Edition location 5666/7240.

39. "Science: The Innoculation Against Charlatans," *Star Talk* (YouTube channel), https://www.youtube.com/watch?v=6Lc_r-Me58M.

40. "Modern Science Documents the Maharishi Effect," Maharishi Programmes, https://maharishi-programmes.globalgoodnews.com/maharishi-effect/research.html. What TM proponents and promotional materials don't tell us is that most of these studies are unpublished, self-published, or in obscure journals. Nor are they true replications as each uses different combinations of social indicators. As far as I know, TM researchers have never adopted stronger research designs or abided by ethical standards of data transparency in response to valid informed criticism.

Chapter 13: The Power of the Pyramid

1. "The Myth the MythBusters Refused to Touch," *Adam Savage's Tested* (YouTube channel), November 9, 2024:(00:00–00:46), https://www.youtube.com/watch?v=nYHi8PLjczw.

2. I encourage readers to do their own online searching for healing pyramids. You'll find they're made from all kinds of materials and in a wide range of sizes, from wearable on the wrist to capable of seating an entire family.

3. Cindy Rose, "Pyramid Power: Many Believe It Has the Angles," *The Philadelphia Inquirer*, February 15, 1977.

4. Edwin Newman, "O Great Cheops, What Bath Thy Offspring Wrought?," *The New York Times*, August 29, 1976.

5. "Pyramid Energy Experiment," *Agui007* (YouTube channel), December 14, 2014, https://www.youtube.com/watch?v=uGGBlx46-38.

6. "Do Pyramids Have Strange Powers? 7 Day Experiment Reveals SECRET," *PraveenMohan* (YouTube channel), December 2, 2019, https://www.youtube.com/watch?v=ExX65rG7FQg&t=364s.

NOTES

7. "Egypt's 'History of Humanity' Monuments Face Climate Change Threat," *The Jerusalem Post*, December 20, 2019, https://www.jpost.com/middle-east/egypts-history-of-humanity-monuments-face-climate-change-threat-611584.

8. Erich van Däniken, *Chariots of the Gods?* (Souvenir Press Ltd., 1969).

9. "Pyramid Architecture," Britannica, accessed March 27, 2025, https://www.britannica.com/technology/pyramid-architecture.

10. This specification is based on the location of the pharaoh's burial chamber in the Great Pyramid of Cheops.

11. This possibility was mentioned on the *Mythbusters* TV program. Pyramid power was put to the test in the 2005 Season with "Jet Pack," which aired June 9, 2005.

12. Again, this experiment predated Praveen's by years. If he read about our project and emulated its three conditions, he never let on. But it easily could have been a coincidence. It would have been daunting to build granite versions of more complex polyhedrons beyond the four-sided tetrahedron (pyramid), five-sided prism (tent), or six-sided cube (box).

13. "Social Impact Theory," *Encyclopedia of Social Psychology*, ed. Roy F. Baumeister and Kathleen D. Vohs (Sage, 2007):904–5, doi:https://doi.org/10.4135/9781412956253.n534.

14. We used a rather elaborate method involving bunches of bananas and a panel of judges to arrive at matched pairs to be used for experimental sessions. To eliminate the possibility that one banana really was fresher than the other, we alternated their alleged container for each new subject.

15. For example, see "Paranormal America 2018," Chapman University, October 16, 2018, https://blogs.chapman.edu/wilkinson/2018/10/16/paranormal-america-2018/.

16. Barry Markovsky and Shane R. Thye, "Social Influence on Paranormal Beliefs," *Sociological Perspectives* 44, no. 1 (2001):21–44; "Social Influence and the Power of the Pyramid," *Skeptic* 9, no. 3 (2002):36–41.

Chapter 14: ESP: Can It Be?

1. Robin Williams, quoted at Goodreads, https://www.goodreads.com/quotes/120432-if-it-s-the-psychic-network-why-do-they-need-a.

2. It's too bad these polls aren't more nuanced. I'd count myself in the "open to the possibility" category. But this doesn't differentiate between a wide range of people. If I had to put a number on my belief in the likelihood ESP exists as claimed, it probably would be around 1 or 2 percent. But a 99-percent believer would fall in the same polling category as me. My sense—based solely on personal impressions from lots of reading and lots of talking to other people about ESP—is that the majority of "opens" lean more toward believing than disbelieving. For some typical polls, see Tom W. Rice, "Believe It or Not: Religious and Other Paranormal Beliefs in the United States," *Journal for the Scientific Study of Religion* 42, no. 1 (2003):95–106, https://doi.org/10.1111/1468-5906.00163; and David W. Moore, "Three in Four Americans Believe in Paranormal," *Gallup News Service*, June 16, 2005, https://news.gallup.com/poll/16915/Three-Four-Americans-Believe-Paranormal.aspx.

3. The Parapsychological Association is the leading organization of its kind and has around three hundred members. See "Parapsychological Association," Wikipedia, accessed March 27, 2025, https://en.wikipedia.org/wiki/Parapsychological_Association.

NOTES

4. Ray Hyman, *The Illusive Quarry: A Scientific Appraisal of Psychical Research* (Prometheus Books, 1989); James Alcock, *Parapsychology, Science or Magic* (Pergamon Press, 1981); and Paul Kurtz (ed.), *A Skeptics Handbook of Parapsychology* (Prometheus Press, 1985).

5. "Peter Venkman's ESP Test," *Ghostbusters* (YouTube channel), November 19, 2020, https://www.youtube.com/watch?v=HW8Ua49dCYk.

6. This paragraph assumes that both sides are talking about the same thing. That's not always true. Sometimes it turns out that the disagreement stems from the sides holding different meanings for the same words. For example, an atheist may declare "There's no god," where in her mind "god" refers to an omnipotent, omniscient, at times temperamental entity described in some very popular religious texts. On the other hand, a believer may declare "There's definitely a god," but in his mind, that means "Nature, love, and the physical laws of the universe." The atheist may have found this version of god totally unobjectionable had it been defined at the start. The lesson: Define your terms.

7. I should point out that parapsychologists rarely use the term "ESP." When speaking generically about psychic phenomena, they most often use the term "psi," as in the Greek letter pronounced "sigh." I'll continue to use the more familiar of the two terms.

8. Critics also discovered more subtle and complex problems involving experimenter biases, statistical tomfoolery, and fraud.

9. Arthur S. Reber and James E. Alcock, "Searching for the Impossible: Parapsychology's Elusive Quest," *American Psychologist* 75, no. 3 (2020):391–99, http://dx.doi.org/10.1037/amp0000486.

10. "Remote Viewing," The Skeptic's Dictionary, October 27, 2015, https://www.skepdic.com/remotevw.html.

11. Benjamin Radford, "Playing With Past Lives: The Virginia Boy and the Dead Marine," *Skeptical Inquirer* 39, no. 5 (2015):33, https://skepticalinquirer.org/2015/09/playing-with-past-lives-the-virginia-boy-and-the-dead-marine/.

12. Leonard Angel, "Empirical Evidence for Reincarnation? Examining Stevenson's 'Most Impressive' Case," *Skeptical Inquirer* 18, no. 5 (1994):481–87, https://skepticalinquirer.org/1994/09/empirical-evidence-for-reincarnation-examining-stevensons-most-impressive-case/.

13. "Psychic Staring Effect (Scopaesthesia)," The Skeptic's Dictionary, August 14, 2011, https://skepdic.com/staringeffect.html.

14. Daryl J. Bem, "Feeling the Future: Experimental Evidence for Anomalous Retroactive Influences on Cognition and Affect," *Journal of Personality and Social Psychology* 100, no. 3 (2011):407–25, https://doi.org/10.1037/a0021524.

15. Stuart J. Ritchie et al., "Failing the Future: Three Unsuccessful Attempts to Replicate Bem's 'Retroactive Facilitation of Recall' Effect," *PLOS One* 7, no. 3 (2012) https://doi.org/10.1371/journal.pone.0033423.

16. Reber and Alcock, "Searching for the Impossible," 391.

17. Susan J. Blackmore, *In Search of the Light: The Adventures of a Parapsychologist* (Prometheus Books, 1996).

18. Ray Hyman, "Cold Reading: How to Convince Strangers That You Know All About Them," *The Zetetic* 1, no. 2 (1977):18–37, https://skepticalinquirer.org/1977/04/cold-reading-how-to-convince-strangers-that-you-know-all-about-them/.

19. Barry Markovsky, "Paranormal Professionals," Annual Meeting of the Society For Scientific Exploration, Las Vegas, 1997.

20. John Scarne, *Scarne's Tricks* (Crown Publishers, 1950).

21. This was a central thesis of Thomas S. Kuhn's *The Structure of Scientific Revolutions, 4th Edition* (University of Chicago Press, 2012).

22. Hyman, 1977. Also, see Kari Coleman, "My Psychic Adventure," Quackwatch, September 22, 2020, https://quackwatch.org/my-psychic-adventure/. The author gives a wonderful first-person account of how easy it was to learn and then successfully use standard psychic techniques—without having a shred of psychic abilities.

23. I recommend Chris Ramsay's short instructional video on cold reading techniques, including analyses of some famous psychics at work. See "These 'Psychics' Are Using Magic Tricks," *Chris Ramsay* (YouTube channel), June 8, 2020, https://www.youtube.com/watch?v=yialEM_T2M0.

24. C.Y. Nicholson et al., "The Role of Interpersonal Liking in Building Trust in Long-Term Channel Relationships," *Journal of the Academy of Marketing Science* 29, no. 1 (2001):3–15, https://doi.org/10.1177/0092070301291001.

25. The brain and most cellular tissue will live a while beyond this moment, so "death" can be defined other ways. See Christopher A. Pallis, "Death: Process or Event," Britannica, accessed March 27, 2025, https://www.britannica.com/science/death/Death-process-or-event.

26. There are high-priced psychic scammers who pose real financial dangers to repeat clients. It's certainly possible that the psychic fair psychics were trolling for new customers. *Caveat emptor.* See "Psychic Scams," AARP, April 28, 2022, https://www.aarp.org/money/scams-fraud/psychic/.

27. Hilary George-Parkin, "When Is Fortune-Telling a Crime?," *The Atlantic*, November 14, 2014, https://www.theatlantic.com/business/archive/2014/11/when-is-fortune-telling-a-crime/382738/.

CHAPTER 15: ARE WE *REAL* DOCTORS?

1. Terry Pratchett, "Back in Black," British Broadcasting Corporation, 2017, https://www.imdb.com/title/tt6468114/.

2. If you go online to find the correct pronunciation for "naturopathy" and "naturopathic," you'll find multiple versions for each. Marcia pronounced them "na-chur-AW-pa-thee" and "na-chur-oh-PATH-ic." But your internal voice can pronounce them any way it wishes.

3. "Naturopathy," National Center for Complimentary and Integrative Health, September 2017, https://www.nccih.nih.gov/health/naturopathy.

4. Edzard Ernst, "How Much of CAM [Complementary and Alternative Medicine] is Based on Research Evidence?," *Evidence-Based Complementary and Alternative Medicine*, June 18, 2011, https://doi.org/10.1093/ecam/nep044; "Acupuncture: A Point in the Right Direction, or a Stab in the Dark?," Harvard Health Publishing, May 3, 2017, https://www.health.harvard.edu/blog/acupuncture-a-point-in-the-right-direction-or-a-stab-in-the-dark-2017050311672; Charlotte M. Kohn and Priyamvada Paudyal, "A Systematic

NOTES

Review and Meta-Analysis of Complementary and Alternative Medicine in Asthma," *European Respiratory Review* 26, January 31, 2017, doi:10.1183/16000617.0092-2016.

5. "Principles of Naturopathic Medicine," American Association of Naturopathic Physicians, 2011, https://naturopathic.org/page/PrinciplesNaturopathicMedicine.

6. Ian Coulter et al., "Vitalism—A Worldview Revisited: A Critique of Vitalism and Its Implications for Integrative Medicine," *Integrative Medicine: A Clinician's Journal* 18, no. 3 (2019):60–73, PMID: 32549817; PMCID: PMC7217401.

7. Stephen Barrett, "A Close Look at Naturopathy," Quackwatch, November 26, 2013, https://quackwatch.org/related/naturopathy/.

8. https://naturemed.org/.

9. "Naturopathy," Wikipedia, accessed March 28, 2025, https://en.wikipedia.org/wiki/Naturopathy.

10. "Myth-Busting the Naturopathy Wikipedia Page," American Association of Naturopathic Physicians, accessed March 28, 2025, https://naturopathic.org/page/FalseWiki.

11. The Naturowatch site provides a comprehensive "skeptical guide to naturopathic history, theories, and practices." See https://quackwatch.org/naturopathy/.

12. Dugald Seely et al., "Naturopathic Medicine for the Prevention of Cardiovascular Disease: A Randomized Clinical Trial," *Canadian Medical Association Journal* 185, no. 9 (2013):E409–16, https://doi.org/10.1503/cmaj.120567.

13. Marina DeWit, "Homeopathy Industry Threatened by New FDA Guidance," U.S. Small Business Administration, January 27, 2020, https://advocacy.sba.gov/2020/01/27/homeopathy-industry-threatened-by-new-fda-guidance/.

14. For example, see "Homeopathic Dilution," Nelsons, accessed March 28, 2025, https://www.nelsons.net/en-us/footer/homeopathic-dilution/.

15. This practice evokes the *fallacy of false precision*. There's no chemical basis for the exacting procedures. Their main function, I suspect, is to inspire undue confidence in practitioners and patients.

16. "Homeopathic Products," U.S. Food and Drug Administration, accessed March 28, 2025, https://www.fda.gov/drugs/information-drug-class/homeopathic-products.

17. Many homeopathic treatments now come diluted in other inert forms such as gels, creams, and tablets.

18. Erin Ross, "Time for Homeopathic Remedies to Prove That They Work?," National Public Radio, December 2, 2016, https://www.npr.org/sections/health-shots/2016/12/02/504004506/time-for-homeopathic-remedies-to-prove-that-they.work.

19. "Homeopathy, Quackery and Fraud | James Randi | TED," *TED* (YouTube channel), April 20, 2010, https://www.youtube.com/watch?v=c0Z7KeNCi7g&t=682s.

20. Martin Robbins, "A Challenge to Homeopaths: How Does One Overdose?," *The Guardian*, November 21, 2010, https://www.theguardian.com/science/the-lay-scientist/2010/nov/21/1.

21. Dana Ullman, "The Case FOR Homeopathic Medicine: The Historical and Scientific Evidence," *Huffington Post*, December 6, 2017, https://www.huffpost.com/entry/the-case-for-homeopathic_b_451187.

22. https://clinicaltrials.gov/ct2/resources/trends.

23. In researching this chapter, I learned that Ullman has built a reputation as an avid and tireless advocate for homeopathy who routinely misrepresents prior research. Addi-

NOTES

tionally—as I describe in another note to follow—I've personally experienced his use of argument fallacies, rhetorical slides, red herrings, and hilariously vicious personal insults. See "Dana Ullman," Rational Wiki, accessed March 28, 2025, https://rationalwiki.org/wiki/Dana_Ullman.

24. The drug was betahistine. A 2016 article concluded that any evidence of its effectiveness should be treated with caution as much of it comes from low-quality research. See https://doi.org/10.1002/14651858.CD010696.pub2.

25. Thanks to my hiking buddy and ENT specialist Dr. Mark Montgomery for this insight. Also see Louisa Murdin et al., "Betahistine for Symptoms of Vertigo," *Cochrane Database of Systematic Reviews*, June 21, no. 6, 2016, https://doi.org/10.1002/14651858.CD010696.pub2; and "Vertigo," National Health Service (Scotland), accessed March 25, 2025, https://www.nhsinform.scot/illnesses-and-conditions/ears-nose-and-throat/vertigo.

26. I reached out to Dana Ullman to give him the chance to defend his article. His reply did four things:

   1. He lashed out at me with personal insults and admonishments—that as a sociologist I don't know enough about homeopathy; as a skeptic I'm arrogant and my mind is closed; and that I "cherry-picked" the quotes, dishonestly taking them out of context from the articles he cited. (For the record, neither my original message to Ullman, nor my reply to his said anything at all about him personally. I noted as much in my response to him. His next reply was angrier and more insulting than his first by about ten times and was equally uninformative.)
   2. With unintended irony, Ullman's reply cherry-picked quotes from the same articles—but only those that expressed optimism (*not* clear support) on behalf of some very specific findings. He didn't address the summary disclaimers and cautions that I quoted, nor why the authors would have expressed such caution if the results were indeed so clearly supportive.
   3. He sent me links to even more articles (mostly his own) that he believes support homeopathy's claims.
   4. He raised a number of points that were irrelevant, false, or both. For instance, he claimed that you can't evaluate any one homeopathic treatment, only the whole system. That's just not true—as exemplified by the articles he cited showing research on individual treatments. (For discussion of this and other fallacies, see Edzard Ernst, "Thirteen Follies and Fallacies About Alternative Medicine," *EMBO Reports* 14, no. 12 (2013):1025–26, https://doi.org/10.1038/embor.2013.174.)

The bottom line is that Ullman's article made false and exaggerated statements about the clinical evidence for homeopathy, and his shotgun rebuttal hit everything but the target. An excellent summary of his ineffective defensive strategies can be found under "Dana Ullman" at the Rational Wiki page cited earlier, https://rationalwiki.org/wiki/Dana_Ullman.

27. Apologies to Zoë Blaylock from my writing group. She was not a fan of the turtle imagery because she loves turtles. Be assured, my friend, that the imaginary vehicle was righted by an imaginary tow truck and brought in for successful repairs. Long may it run.

Chapter 16: That's the Spirit!

1. Stephen King, *The Shining* (Doubleday, 1977).

2. My pseudonym for the shop owner is a nod to Benjamin Radford, paranormal investigator extraordinaire and author of a number of books that are staples on skeptics' shelves. He was responsive and supportive when I had questions while working on an article about this case for *Skeptic* magazine.

3. Ballard, Jamie, "45% of Americans Believe That Ghosts and Demons Exist," YouGov, October 21, 2019, https://today.yougov.com/society/articles/25908-paranormal-beliefs-ghosts-demons-poll; also, this IPSOS poll: "Majority of Americans Believe in Ghosts (57%) and UFOs (52%)," October 31, 2008, https://www.ipsos.com/en-us/news-polls/majority-americans-believe-ghosts-57-and-ufos-52.

4. "YouGov/The Sun Survey Results," YouGov, poll conducted October 2014, accessed March 28, 2025, https://ygo-assets-websites-editorial-emea.yougov.net/documents/Sun-Results_141027_Ghosts-Website_zNT96VG.pdf

5. "Paranormal Television," Wikipedia, accessed March 28, 2025, https://en.wikipedia.org/wiki/Paranormal_television.

6. See, for example, Benjamin Radford, *Scientific Paranormal Investigation* (Rhombus Publishing Co., 2010); Benjamin Radford, "Ghost-Hunting Mistakes," *Skeptical Inquirer* 34, no. 6 (2010):44–47; Sharon A. Hill, *Scientifical Americans* (McFarland & Co. Publishers, 2017), chapter 3.

7. More details can be found here: Barry Markovsky, "Notes on a Haunting," *Skeptic* 26, no. 2 (2021):37–45.

8. Ben used Nest IQ indoor cameras. These switch automatically between day and night mode. That is, when lighting is low, they record using their own infrared lights that surround the camera lens.

9. "Why Do I See Orbs or Bubbles When My Camera Is Using Night Vision?," SimpliSafe Support, accessed March 28, 2025, https://support.simplisafe.com/articles/cameras/why-do-i-see-orbs-or-bubbles-when-my-camera-is-using-night-vision/634492a5d9a8b404da76cccb; also, "Paranormal 101 | Orbs Explained | Paranormal Investigators of Milwaukee," *Paranormal Milwaukee* (YouTube channel), November 27, 2019, https://www.youtube.com/watch?v=QnaTT_MkbUg.

10. Claire Elliott, "Meaning in Randomness," The Psychologist, The British Psychological Society, March 7, 2018, https://thepsychologist.bps.org.uk/volume-31/april-2018/meaning-randomness. Also, in his TED talk, Michael Shermer beautifully demonstrates how these effects color what we hear in a song played backward: "Why People Believe Weird Things | Michael Shermer," *TED* (YouTube channel), April 15, 2008 (starting at 9:05), https://www.youtube.com/watch?v=8T_jwq9ph8k&t=52s.

11. "Re-creating the 'Flying Bible Incident' From a Ghost Documentary," *Barry Markovsky* (YouTube channel), June 5, 2022, https://www.youtube.com/watch?v=R2Pk-Qc-Jzo4&t=2s.

Chapter 17: Numbers Game

1. John Saxon, untitled, in "Mathematical Poems," *Discover Magazine*, November 5, 2005, https://www.discovermagazine.com/the-sciences/mathematical-poems.

2. Some of the material that follows originally appeared in this article: Barry Markovsky, "Happy Twosday! Why Numbers Like 2/22/22 Have Been Too Fascinating for Over 2,000 Years," *The Conversation*, February 17, 2022, https://theconversation.com

NOTES

/happy-twosday-why-numbers-like-2-22-22-have-been-too-fascinating-for-over-2-000-years-176093.

3. One such list can be found here: David Mikkelson, "Are These 'Coincidences' Linking Lincoln to Kennedy Real?," Snopes, June 11, 1999, https://www.snopes.com/fact-check/linkin-kennedy/.

4. "Numerology," The Skeptic's Dictionary, October 27, 2015, https://skepdic.com/numology.html.

5. Aliza Kelly, "Numerology Numbers: What Is Your Life Path Number?," *Allure*, December 1, 2021, https://www.allure.com/story/numerology-how-to-calculate-life-path-destiny-number.

6. Matt Beech, "Destiny Number 5—Embracing Freedom and Adventure," accessed March 29, 2025, https://mattbeech.com/numerology/destiny-number/destiny-5/.

7. Michael Drosnin, *The Bible Code* (Simon & Schuster, 1998).

8. David E. Thomas, "Hidden Messages and *The Bible Code*," *Skeptical Inquirer* 21, no. 6 (1997): 30–36, https://skepticalinquirer.org/1997/11/hidden-messages-and-the-bible-code.

9. Allyn Jackson, "*The Bible Code*" book review, *Notices of the American Mathematical Society* 44, no. 8 (1997): 935–38, https://www.ams.org/notices/199708/review-allyn.pdf.

10. Jonathan Freedland, "The Assassination of Yitzhak Rabin: 'He Never Knew It Was One of His People Who Shot Him in the Back,'" *The Guardian*, October 31, 2020, https://www.theguardian.com/world/2020/oct/31/assassination-yitzhak-rabin-never-knew-his-people-shot-him-in-back.

11. Jerome H. Barkow et al. (eds.), *The Adapted Mind* (Oxford University Press, 1992).

12. Jeff Grabmeier, "This Is Your Brain Detecting Patterns," *Science Daily*, May 31, 2018, https://www.sciencedaily.com/releases/2018/05/180531114642.htm.

13. "Dopamine, Facial Recognition Connected, Study Shows," *Science Daily*, October 3, 2016, https://www.sciencedaily.com/releases/2016/10/161003093240.htm.

14. Laura Ferreri et al., "Dopamine Modulates the Reward Experiences Elicited by Music," *Proceedings of the National Academy of Sciences* 116, no. 9 (2019): 3793–98, https://doi.org/10.1073/pnas.1811878116.

15. Hope Cristol et al., "Dopamine: What It Is & What It Does," WebMD, July 9, 2024, https://www.webmd.com/mental-health/what-is-dopamine.

16. John W. Hoopes, "11-11-11, Apophenia, and the Meaning of Life," *Psychology Today*, November 11, 2011, https://www.psychologytoday.com/us/blog/reality-check/201111/11-11-11-apophenia-and-the-meaning-life.

17. Michael Shermer, "Patternicity: Finding Meaningful Patterns in Meaningless Noise," *Scientific American* 299, no. 6 (2008), https://www.scientificamerican.com/article/patternicity-finding-meaningful-patterns/.

18. Barry Markovsky, "How Did the Superstition That Broken Mirrors Cause Bad Luck Start and Why Does It Still Exist?," The Conversation, June 29, 2021, https://theconversation.com/how-did-the-superstition-that-broken-mirrors-cause-bad-luck-start-and-why-does-it-still-exist-162889.

19. Alethia Luna, "11 11 Meaning: Do You Keep Seeing This Unusual Number?," Lonerwolf, July 25, 2024, https://lonerwolf.com/1111-meaning/.

20. Joseph Jastrow, "Science and Numerology," *The Scientific Monthly* 37, no. 5 (1933):448–50, https://www.jstor.org/stable/15626.

21. Tracy V. Wilson, "How Numerology Works," How Stuff Works, December 15, 2022, https://science.howstuffworks.com/science-vs-myth/extrasensory-perceptions/numerology.htm.

22. Meghan Holohan, "'In Disbelief': Mom Gives Birth to 3 Girls on the Same Day Exactly 3 Years Apart," *Today*, October 20, 2021, https://on.today.com/3KlJ3ea.

23. "Derren Brown–10 Heads in a Row," *ThinkSceptically* (YouTube channel), April 8, 2012, https://www.youtube.com/watch?v=XzYLHOX50Bc.

24. The actual formula is $½^{10}$ = 1/1,024 = 0.000977.

25. This calculation is for how many flips to expect until getting ten heads in a row. It's not the same as the calculation for getting ten heads in a row with ten flips of a coin. See "The Derren Brown Coin Flipping Scam," nrich, accessed March 29, 2025, https://nrich.maths.org/6954/solution.

26. Brown was not quite so lucky. In another video, he says that it took a nine-hour recording session—possibly an exaggeration but within the realm of possibility. As for video magic, there was indeed a cut between Brown's announcing what he was about to do and the first flip of the coin: a stylized, slow-motion insert of a coin flip. See "Derren Brown—10 Heads in a Row (Explanation)," *ThinkSceptically* (YouTube channel), April 8, 2012, https://youtu.be/n1SJ-Tn3bcQ?t=70.

27. David E. Thomas noted that Drosnin used a skip code of around 33,000 to predict that incumbent Bill Clinton would win the next US presidential election. See Ben Radford and Celestia Ward, "The Bible Code Resurfaces, with David Thomas," *Squaring the Strange* podcast (24:42), February 24, 2023, https://squaringthestrange.libsyn.com/episode-195-the-bible-code-resurfaces-with-david-thomas.

28. David E. Thomas, "Hidden Messages."

29. Brendan McKay, "Assassinations Foretold in Moby Dick!," accessed March 29, 2025, http://users.cecs.anu.edu.au/~bdm/dilugim/moby.html.

30. Yes.

Chapter 18: Circling the Square

1. "Paranoid Personality Disorder," PsychDB, January 27, 2024, https://www.psychdb.com/personality/paranoid.

2. Steven Pinker, *Enlightenment Now: The Case for Reason, Science, Humanism, and Progress* (Penguin, 2018), 16–19.

3. Jan-Willem van Prooijen and Mark van Vugt, "Conspiracy Theories: Evolved Functions and Psychological Mechanisms," *Perspectives on Psychological Science* 13, no. 6 (2018):770–88, https://doi.org/10.1177/1745691618774270.

4. Mark R. Cheatham, "Conspiracy Theories Abounded in 19th-Century American Politics," *Smithsonian Magazine*, April 11, 2019, https://bit.ly/3JaVqvW.

5. Many readers will remember that, as part of the al-Qaeda conspiracy, another jet crashed into the Pentagon, and a fourth one heading to DC was hijacked by passengers and crashed in a field in Shanksville, Pennsylvania. Here and later, I focus only on the World Trade Center crashes for the sake of brevity.

NOTES

6. Michael Shermer, *Conspiracy: Why the Rational Believe the Irrational* (Johns Hopkins Press, 2022); Mick West, *Escaping the Rabbit Hole: How to Debunk Conspiracy Theories Using Facts, Logic, and Respect* (Skyhorse Publishing, 2018).

7. "Square the Circle," Dictionary.com (Idioms and Phrases), https://www.dictionary.com/browse/square-the-circle.

8. Antonio Damasio, *Descartes' Error: Emotion, Reason, and the Human Brain* (Penguin Books, 2005).

9. Mary C. Lamia, "What Are Beliefs?," Psychology Today, September 20, 2021, https://www.psychologytoday.com/us/blog/intense-emotions-and-strong-feelings/202109/what-are-beliefs.

10. An excellent nontechnical review of the evidence was published by Phil Molé in *Skeptic* 12, no. 4 (2006):30–42. For more technical treatments, I'd recommend these: "World Trade Center Investigation," National Institute of Standards and Technology, accessed March 29, 2025, https://www.nist.gov/world-trade-center-investigation; Zdenĕk P. Bažant and Mathieu Verdure, "Mechanics of Progressive Collapse: Learning From World Trade Center and Building Demolitions," *Journal of Engineering Mechanics* 133, no. 3 (2007):308–19, https://doi.org/10.1061/(ASCE)0733-9399(2007)133:3(308).

11. Thomas W. Eager and Christopher Musso, "Why Did the World Trade Center Collapse? Science, Engineering, and Speculation," *JOM* 53, no. 12 (2001):8–11, https://www.tms.org/pubs/journals/jom/0112/eagar/eagar-0112.html.

12. "Tallest Building Demolitions in the World," *MegaBuilds* (YouTube channel), August 15, 2021, https://www.youtube.com/watch?v=Mb4MNe1vk70.

13. "Ronan Point," Wikipedia, accessed March 29, 2025, https://en.wikipedia.org/wiki/Ronan_Point.

14. West, *Escaping the Rabbit Hole*, 141–42.

15. This fallacy is called "argument from personal incredulity."

16. West, *Escaping the Rabbit Hole*, chapter 2.

17. Shermer, *Conspiracy*, 131–32.

18. Stephen Jay Gould, "Foreword: The Positive Power of Skepticism," in Michael Shermer, *Why People Believe Weird Things* (W.H. Freeman, 1997), x.

19. Although Awake Dating received a lot of attention when it went live in 2016, it appears to be defunct as of this writing. However, Vice News talks about new sites that have filled the void: Tim Hume, "Conspiracy Theorists are Setting Up Their Own Dating Sites," March 14, 2023, https://www.vice.com/en/article/conspiracy-theorist-dating-website/.

20. Dan Evon, "Was COVID Vaccine Microchip Detected With Pet Scanner?," Snopes, July 2, 2021, https://www.snopes.com/fact-check/covid-vaccine-pet-microchip/.

INDEX

AANP. *See* American Association of Naturopathic Physicians
abuse, 86, 136–37, 141
acid (LSD), 18
adaptive conspiracism, 220
adolescence, 27, 42, 60–61, 62, 63, 67
advertising blimp, 50
agenticity, 112, 113
agnosticism, 41
Air Force, United States, 53
airplane, 50, 53; in World Trade Center Tragedy, 229–32
Alcock, James, 165
aliens, 16, 128; abductions, 54, 127; ET, xii, 19, 47–55, 148, 247n3, 247n15; intelligent life, 52, 218; invasion of, 47; pyramid power and, 148; UFO, xi, 46–47, 48–55, 247n3
The All-Iowa Holistic and Psychic Extravaganza!, 155–56
altered states, of consciousness, 57, 62
alternative explanations, x, xi
alternative medical practices, 177–78
American Association of Naturopathic Physicians (AANP), 177–78
American Psychological Association, 62
American Society of Dowsers, 115, 122–23
antisemitism, 34–35
apophenia, 205–6
apparition, xiii, 7–8
Arnold, Kenneth, 48
art, paranormality and, 156
artificial intelligence, 18

assassination, of Kennedy, John, F., 38, 221
astrologers, 97, 100–101, 104, 155
astrology, 16, 95, 113, 128, 252n5; belief in, 97–99, 107; claims of, 99, 100–101; as coping mechanism, 105; evidence for, 97, 99, 100; horoscopes from, 96–97, 100, 103, 105, 253n17; introvert–extrovert test and, 99–100; moment of birth, 101, 104; natal chart, 101–2, 104, 252n13; predictive powers of, xii, 99; science and, 97–98, 99, 105; sun sign, 96, 100, 107; zodiac signs for, 17, 99, 100, 104, 253
astronologers, 98
astronomy, 51, 96, 97, 98, 218
atheist, 41, 260n6
audio mirage, 192–93
auditory pareidolia, 192
aura, 63, 146, 155
autoimmune system, 73, 74, 75
Awake Dating (dating website for conspiracy theorists), 233, 267n19

Balducci Technique, 65–66
baloney-detection kit, 232–33
Barnum, P. T., 107
Barnum Effect, 107
Barrie, J. M., 56
bartering, baseball cards, 209–10
baseball cards, 209–10
the Beatles, 32–33, 130, 255n3
Beecher, Reverend, 35–37, 39–40

beer, 56, 57, 72, 121
beliefs, 227, 228, 235; in astrology, 97–99, 107; of children, 4; conspiratorial, 211; dis-, 31, 41–42; false, 154, 219, 223, 225, 233; in ghosts, 190; magical, 29; popular, 37, 53, 145; religious, x, 75, 86, 195; Shermer on, 142; social impact theory influence on, 89, 151, 153–54; social influences on, 52–53; supernatural, 61, 76, 84, 191, 194, 237; systems for, xii, 79, 105; unshakeable, xi. *See also* prayer
believer, 15, 91, 211; conspiracy theory, 226; ESP, 160, 166; in extraterrestrials, 55; in God, 78, 260n6; non-, 220; prayers and, 75, 77; in pyramid power, 145; science-minded, 76; true, xi, 103, 121
Bem, Daryl, 163–64, 165
Benson, Herbert, 77–78
The Beverly Hillbillies (TV sitcom), 73
biases, xi–xii, 149, 163, 260n8; confirmation, 103, 142; cultural, 137; natural, 170; randomization and, 138–39
Bible, 40, 41, 189; flying, 194, 195, 196; New Testament, 37, 123; numerology in, 203; Old Testament, 35, 36, 203, 206, 208
Bible Code, 202, 203–4, 208
The Bible Code (Drosnin), 203, 204, 208
biblical claims, 20
Biden, Joe, 222
Big Cupid (Russian cabal), 233
bin Laden, Osama, 217
binomial probability, 254n3
birthday paradox, 253n6
Blackmore, Susan, 166
blooming dust, 191, 192
bombing, of World Trade Center, 216–18
born-again experience, 83–84, 86, 89–90, 94
Boston Red Sox (baseball team), 209–10

Boston Strangler, 32
Bradbury, Ray, 24–25
the brain, 19, 261n24; agenticity of, 112, 113; autoimmune system functioning and, 75; fMRI, 18; hallucinogenic drugs effects on, 18; neurons in, 17, 18, 113; optical illusions and, 223–25; paternicity of, 112–13; perception of, 20, 52, 186; soul and, 17–18; TM and, 139, 140, 141. *See also* consciousness
Brain Games, on National Geographic Channel, 64
Broca's Brain (Sagan), 40
Brown, Derrin, 64, 207, 208, 266n26
bubonic plague, 36, 39
burial chamber, of pharaoh, 145, 259n10
Bush, George, W., 217, 218

California Personality Inventory (personality test), 101
card-guessing experiment, 159–60, 161–62
Carlson, Shawn, 100–101
case studies, 183
Catch-22 (Heller), 232
Catholicism, 34, 83, 86
cause-and-effect claims, 76
central intelligence agency (CIA), 163, 213, 217, 218, 227
chants, 58, 59, 67
Chariots of the Gods? (Däniken), 148
Charleston, South Carolina, 185, 186, 196
chemtrails, 221
Cheops pyramid, 146, 259n10
children: beliefs of, 4; dreams of, 6–7; ESP and, 163; human memory in, 3–5; magical thinking in, 29; nightmares of, 9–10, 13–15; psychology for, 27; reincarnation promoted by, 163; religious experiences of, 31–32; socialization and, 27–28
China First (restaurant), 212, 215–16, 227–28

INDEX

Christianity, 34, 188; born again experience in, 83–84, 86, 89; Christmas celebrated in, 22–24, 26, 27–29; skepticism in, 22
Christian Scientists, 70–75, 78–80
Christmas (Christian holiday), 22–24, 26, 27–29
Church, Francis, P., 22
church group, 70–71, 84–86, 87–90
The Church of Christ, Scientist, 73–74, 79
Church of Scientology, 251n4
CIA. *See* central intelligence agency
claims, xiii, 165; of astrology, 99, 100–101; biblical, 20; cause-and-effect, 76; ESP, 159, 165; levitation, 65; of naturopathy, 177–78; paranormal, 61, 145, 195; religious, x, 30–31. *See also* extraordinary claims
clairvoyance, 19–20, 161
Clarke, Arthur, C., 95
climate change, 147
clinical research, 176, 256n12
cognition: motivated, 227; pre-, 20, 161–62, 169, 245n41; re-, 88–89, 209
coincidences, 113, 215; earthquakes and, 110, 111, 112; Lincoln–Kennedy, 201–2, 206–8; more-than-, 206, 207, 208; UFO and, 54; uncanny, x
coin collection, 202
coin flip, 207, 266n25, 266n26
Colasanti, Susane, 60, 61
cold reading, 169–70, 261n22
collective consciousness, 16, 164
collective psyche, 63, 212
CompuScan Project, 105–7
computer monitor, 212, 213
confirmation bias, 103, 142
connecting the dots, 223–25, 234
consciousness, 10, 18, 19, 20; altered states of, 57, 62; collective, 16, 164; unity, 132
consequences, 58; health, 10–11, 141, 222; of illusions, 53; of religious beliefs, xi, 75, 86; social, 86

Conspiracy (Shermer), 222–23
conspiracy theories, 205, 220–23, 227–28, 232; culture and, 226; ESP, 160; QAnon, 22, 234; UFO, 48, 54
conspiracy theorists, 211, 233–34, 267n19
conspiratorial beliefs, 211
Constitution, of United States, 37
constructive skepticism, 143
consumer fraud, in psychic readings, 172–73
Contact (science fiction drama film), 51
context, x, 5, 86–87, 263n26; cultural, 27; emotional, 227; religious, 75; social, 205
controlled demolitions, 218, 222, 223, 228, 229, 230–31
conventional medicine, 71, 175–76, 177, 181
conventional treatment, 179, 182
The Conversation (online news outlet), 198
coping mechanism, astrology as, 105
Copperfield, David, 66
Costner, Kevin, 189
COVID-19 Pandemic, 222, 234
crime rates, 132, 142
cults, 87, 251n4
cultural biases, 137
cultural context, 27
cultural diffusion, 28–29
cultural phenomenon, 54, 55
culture, 11, 19, 237; conspiracy theory and, 226; dreams influenced by, 20; of Jewish people, 35, 40
Curses, the Villain is Foiled (musical theater), 81
customs, 27, 28, 29

Dahl, Roald, 1
Daily Collegian (college newspaper), 95
Dandelion Wine (Bradbury), 24–25
Däniken, Erich van, 148
data set, of Jerusalem study, 136–37
Dave the Psychic, 156–59, 169–73
Dawkins, Richard, 246n8

INDEX

debunking, x, xi, 64, 99, 181, 233
deception, 42, 88, 90, 94; self-, 138, 142, 143; willful, 92
Delta Air Lines Flight 723, 251n8
Dement, William, 6
dementia, 21
demographic categories, 102
Des Moines Area Community College, 134
Destiny Numbers, 202–3, 206, 210
Dickie the Dam and the Big Blue River (Keegan), 68–69, 73
disbelief, 31, 41–42
discernment, 19–20, 23, 86
disclaimers, for homeopathic treatments, 180–81
divination, 206
divine intervention, 76, 92
documentary, on Vital Vapors, 186, 188–89
double-blinding protocol, 124–25, 150–51
dowsers, 115–26
dowsing rods, 116, 117, 118, 121, 126, 127
dowsing tools, 116, 117–18, 125–26
dream analysis, 15–16
dreams, 18, 244n15, 246n43; of children, 6–7; cultures influence on, 20; Dement on, 6; deprivation of, 12; of earthquakes, 109, 110, 111, 112, 113, 253n5; evidence of, 11–12, 16–17; human memory of, 21; lucid, 13–14; nightmares, 9–10, 13–15; night terrors, 16; science on, 16–17; sleep phases and, 7, 11, 12, 19; themes in, 10–11; *Zolar's Encyclopedia and Dictionary of Dreams*, 16
Drosnin, Michael, 203, 204, 208–9, 266n27
dualism, 17, 18, 20
Dubuque, Iowa, 189

earthquake, 108–14, 253n5
East Oregonian (newspaper), 48

Ecstasy (MDMA), 18
Eddy, Mary Baker, 73–74, 78, 79
The Ed Sullivan Show (variety TV show), 32–33
Edward, Mark, 64
efficacy, of homeopathy, 182–84
Egyptian pyramids, 146, 147–48
Egypt's Western desert, 148
Einstein, Albert, 165
electronic voice phenomena (EVP), 187, 192–93, 194
embalming, 148
emotions, 172; conspiracy theories and, 227; context of, 227; human memory and, 3, 26; inflammation impacted by, 75; psychology of, 211
energy products, 156
epistemic corruption, 257n30
equal improvement, 182
Escaping the Rabbit Hole (West), 222
ESP. *See* extra-sensory perception
ET. *See* extraterrestrial being
eternal soul, 71
evidence, xi, 40, 91, 92, 190, 233; for astrology, 97, 99, 100; for Christian Scientists, 74, 75; for conspiracy theories, 218, 220, 221, 226; for dowsers, 125, 127; of dreams, 11–12, 16–17; for ESP, 159, 160, 165, 166, 167, 169; for faith healing, 251n12; for ghosts, 194, 195, 196; of God, 78; for ley lines, 118; magical thinking and, 29; for Maharishi Effect, 133, 134, 137; for naturopathy, 176–79, 180–83, 184; for Ouija board, 64; for pyramid power, 149; repeatable, xii; for Roswell Incident, 48; scientific theory and, 90; for TM, 131, 141; UFO, 49, 51, 52, 53, 54, 55, 247n4; undeniable, 2; for voodoo dolls, 63; of World Trade Center tragedy, 229–32
evolved constructive paranoia, 219–20

INDEX

EVP. *See* electronic voice phenomena
expectancy effects, 192
explosives, 127, 217, 218–19, 228–32, 235
external threats, to social groups, 88
extraordinary claims, xi, xii, 17, 30, 62, 92, 237; of Beecher, 39; conspiracy theories, 211–19; of Vital Vapors, 195
extraordinary phenomenon, x, 128
extra-sensory perception (ESP), xi, xii, 128, 145, 259n2; claims of, 159, 165; clairvoyance, 19–20, 161; evidence for, 159, 160, 165, 166, 167, 169; precognition, 20, 161–62, 169; proponents of, 166, 167–69; psychokinesis, 161, 162; skepticism about, 160–61, 164, 166–69
extraterrestrial being (ET), xii, 19, 47–55, 148, 247n3, 247n15
extraterrestrial hypothesis, 54–55

Fairfield, Iowa, 130, 133, 134, 142
faith, 31, 38–39, 41, 74
The Faith Healers (Randi), 251n12
faith healing, 71, 75, 81, 85, 94, 128, 129, 251n12
Fales, Evan, 136
false beliefs, 154, 219, 223, 225, 233
false conspiracy, 218, 233
false precision, 180, 262n15
falsifiability, science norm of, 102
FDA. *See* Food and Drug Administration
Field of Dreams (movie), 189
finger-tingle method, 117
Flat Earth, 221
flying apparatus, of Copperfield, 66
flying objects, xiii, 48, 193, 194, 195, 196, 197
flying saucer, 48, 51, 130
fMRI. *See* functional magnetic resonance imaging
Food and Drug Administration (FDA), 180
Forer Effect, 107

Fort, Charles, 115
Fort Knox, JABET and, 44–46
fourth responses, ix–x
Framingham, Massachusetts, 199
fruit oxidation, pyramid power and, 147, 149–50, 151, 152, 154, 259n14
functional magnetic resonance imaging (fMRI), 18
The Future is Now! (Bem), 163–64

Garfield family murders, 33–34
Garg, Sachin, 128
Gates, Bill, 222, 234
Gauquelin, Michel, 100
Ghost (movie), 193
Ghostbusters (movie), 159–60, 161–62
ghost-hunting, 189, 190, 192
ghosts, ix, 7–8, 185–86, 188, 189–90, 195
Giza pyramids, 148
god (generic deities), 75–76, 78, 256n9, 260n6
God (Judeo-Christian god), 30, 38, 77–78, 79, 84, 260n6; disbelief in, 41–42; in Old Testament, 35, 203; praying to, 34; skepticism about, 40–41
Godspell (musical theater), 68
golden domes, for meditators, 130–31, 255n5, 255n6
good deeds, 128
Goodman, Linda, 96, 97, 99
Google (internet search engine), xii, 91, 230
Gould, Stephen Jay, 233
Grand Canyon video, of Copperfield, 66, 249n22
"the Granny Effect," 73, 74
The Great Divide, 63
ground zero, World Trade Center tragedy, 230–31
gut feelings, 52, 113

Hahnemann, Samuel, 179–80
hallucinations, ix, 11, 12, 19

INDEX

hallucinogenic drugs, 18
Hanukkah (Jewish holiday), 22–24, 27–29
Hayward Fault, 111
health consequences, 10–11, 141, 222
health promotion counseling, 179
heaven, 32, 34, 38, 41, 42, 79, 105
"Heaven on Earth," 130, 136
Hebrew school, 22, 35, 40–41
Hebrew Torah. *See* Old Testament
Heller, Joseph, 232
Hello, Dolly! (musical theater), 81–82
higher power, 75–76, 86
high school couple, 82–83
hijackers, al-Qaeda, 217, 222, 232, 266n5
Hitler, Adolf, 204, 221
Hodge, David, 77
Holocaust Denial, 221
Holy Spirit, 85
homeopathic remedies, 181, 183
homeopathic treatments, 179–81, 262n17, 263n26
homeopathy, 176, 179–84, 262n17, 262n23, 263n26
homophily, 61
honest liars, 92
horoscopes, 96–97, 100, 103, 105, 253n17
Hovering Monk, 66
human behavior, 86, 88–89
human factors, 52, 151
human memory, 2; in children, 3–5; of dreams, 21; emotions and, 3, 26; fallibility of, 170; false, 3, 62; nightmares and, 10, 14; REM and, 11–12; as a time machine, 25, 26–27
hydrangea flowers, 179–80
Hyman, Ray, 166, 261n21
hyperthymesia, 2
hypnagogic sleep, 7, 11, 19
hypnopompic sleep, 12, 19
hypnosis, 57, 62, 157

ideomotor effect, 64, 126
I Know You're Staring at Me! (Sheldrake), 163

illnesses, 18, 79, 88; homeopathy and, 176, 179; as illusions, 73–74; prayer treating, 71, 73–74; pyramid power and, 145
Illuminati, 234
illusions, 55; illness as, 73–74; nose trick, 1–3, 4–5; optical, 50, 53, 223–25
Indiana Jones and the Temple of Doom (film), 62–63
Indiana Jones franchise, 63
inflammation, impacted by emotions, 75
informed consent, 124, 140–41
infrared cameras, 186, 190, 191, 264n8
In Search of the Light (Blackmore), 166
insomnia, 10–11
Institute for Natural Medicine, 178–79
Institutional Review Board, 141
institutions, xii, 27, 139, 141, 226, 237
intelligent life, 52, 218
intercessory prayers, 76–77
internal threats, to social groups, 88
introvert–extrovert test, 99–100
invisibility, 128, 130, 131
Iowa Academy of Science, 129, 131, 255n7

JABET. *See* Jeff, Andy, Barry, Eddie, and Terry
Jakobsen, Vernon, 115–25, 127, 254n3
Jeff, Andy, Barry, Eddie, and Terry (JABET), 43–46
Jerusalem study, TM and, *131*, 132, 135–42, 257n25
Jesus Christ, 23, 71, 83–85, 93, 94, 123; Christian Scientists and, 74; figurine of, 70; name in vain of, 24, 33; WWJD, 92
Jewish people: antisemitism towards, 34–35; culture of, 35, 40; Hanukkah, 22–24, 27–29; Hebrew school, 22, 35, 40–41; Jerusalem study and, 135, 136
Johns Hopkins University, 256n12
juries, 89

INDEX

Keegan, Tom, 68–69
Kennedy, John, F., 36, 38, 201–2, 204, 206–9, 221
King, Martin Luther, 46
King, Stephen, 185

The Lathe of Heaven (Le Guin), 112
Law of Infinitesimals (Hahnemann), 179
Law of Like Cures Like (Hahnemann), 179
Leavitt, Lindsey, 60
Lebanon War, 136
leg-growing stunt, 84–85, 90–91, 92–94
legitimacy theory, 78
Le Guin, Ursula K., 112
Lennon, John, 32
lens flares, 192
levitation, 58–59, 62, 65–67, 130
ley lines, 118–19, 120, 121, 123, 125
Light as a feather, stiff as a board (paranormal game), 58–59, 65, 67
Lincoln, Abraham, 201–2, 206–9
Lincoln–Kennedy coincidences, 201–2, 206–8
The Lord's Prayer, 31–32, 37
love-bombing, 87
LSD. *See* acid
lucid dreams, 13–14
Lunker Land, 129

magical beliefs, 29
magical thinking, 29
magicians, 25, 66, 92, 166, 168, 173
magic tricks, 92, 194–95
Maharishi Effect, 132, 133, 134, 135, 137, 138–43, 256n13, 258n40
Maharishi International University (MIU), 130–31, 132–33, 134, 138, 255n6, 257n25
Maharishi Mahesh Yogi, 129–30, 132, 135, 255n3
Mars, 48–49, 100
Mars effect (Gauquelin), 100
Matthew, New Testament Book of, 37
McKay, Brendan, 209

MDMA. *See* Ecstasy
medical treatment, 74–75, 80, 178–79
medicine: alternative medical practices, 177–78; conventional, 71, 175–76, 177, 181; modern, 74
meditation groups, TM, *131*, 132, 135, 136, 139, 256n13
meditators, 10, 71; golden domes for, 130–31, 255n5, 255n6; mindfulness, 256n12; non-, 130, 133, 139; Siddhi, 135, 256n19; weather effected by, 131, 133–34. *See also* Transcendental Meditation
Melville, Herman, 209
mental picture, 5, 200
mental practices, 246n44
Mercer, Leigh, 197
mindfulness meditation, 256n12
miracles, x, 61, 84–85, 89, 90–91, 93, 251n8
misperceptions, 3, 53, 54, 167, 168
MIU. *See* Maharishi International University
Moby Dick (Melville), 209
modern medicine, as inferior to prayer, 74
Mohan, Praveen, 146–47, 149–50, 259n12
molten steel, 219, 232
moment of birth, 101, 104
monists, 17–18
monsters, 8–9, 13, 14, 20, 21, 49, 185, 201
Moon, 98, 101, 103, 104, 105, 221
Moon Landings, United States, 221
more-than-coincidence, 206, 207, 208
motivated cognition, 227
MUFON. *See* Mutual UFO Network
mummies, 148, 149
Murray, Bill, 159–60
Mutual UFO Network (MUFON), 54, 248n17
Mythbusters (TV program), 259n11

nanny cam, 213
nanothermite, 219, 231, 232
NASA, 51, 221, 232

INDEX

natal chart, 101–2, 104, 252n13
Natick, Massachusetts, 198–99
National Council for Geocosmic Research, 100–101
National Geographic Channel, 64
National Institutes of Health, 138
National Intelligence, Preliminary Assessment from, 54
National Weather Service, 134
natural bias, 170
Nature (journal), 100–101
naturopathic physician, 175–77
naturopathic treatments, 178–79
naturopathic websites, 178
naturopathy, 175–79, 261n2, 262n11
Naturowatch (website), 262n11
Nazis, 204
neural network, 17, 18
neurons, in the brain, 17, 18, 113
Newman, Edwin, 146
new members, of church groups, 87, 88
New Testament, 37, 123
Newton, Isaac, 165
New York Times (newspaper), 130
New York Times Magazine, 146
nightmares, 9–10, 13–15
night terrors, 16
9/11 Truthers, 217–18, 222, 229, 233
nonbeliever, 220
nonlinearity of time, 164
non-meditators, 130, 133, 139
nonverbal cues, 171
norms, 27, 28, 61
nose trick, 1–3, 4–5
number patterns, 205–6, 215
numerical facts, 202, 208
numerical thinking, 197–98, 201–2, 205
numerology, 145, 202, 203–7, 209, 210

Occam's Razor, 247n5
Old Testament, 35, 36, 203, 206, 208
The Onion (satirical newspaper), 108
optical illusions, 50, 53, 223–25
orbs (blobs of light), 187, 189, 190, 191, 192

Orme-Johnson, David, 137, 256n8
Ouija board (paranormal game), 58, 62, 63–65, 255n10
outsiders, 87, 220

pagan rituals, 24
palm reading, 166, 206
parallel universe, 18
paranoia, 32, 213–15, 216, 219–20, 232, 233
paranormal investigator, 194, 263n2
paranormality, 144; art and, 156; claims of, 61, 145, 195; experiences of, 55; experts in, 188; games involved in, 57–60, 62, 63–65, 255n10; services of, 155; YouTube channel on, 65
The Parapsychological Association, 259n3
parapsychologists, 160–61, 165–68, 169, 260n7
parapsychology, 162, 163, 165, 166, 167, 168, 169
Park, Robert, L., 142
parties, adolescent, 56–57, 62, 63, 67
Pascal's Wager, 41–42
patternicity, 112–13
peer groups, 28, 61–62, 67
pendulum, 116, 126
pentagram, 118
perceived threats, 219–20
perceptions, 224, 226–27; of the brain, 20, 52, 186; mis-, 3, 53, 54, 167, 168; tricks of, 65–66. *See also* extra-sensory perception
peripheral vision, ix–x, xiii
personality tests, 99, 101, 102, 105, 106
Personal Traits Checklist, 102
personal transformation, 86
pharaohs, 145, 259n10
phenomenon, 12, 125; cultural, 54, 55; extraordinary, x, 128; of Maharishi Effect, 133; psychic, 161; supernatural, 91, 151; UAP, 247n3
The Philadelphia Inquirer (newspaper), 145
Philips, Emo, 30

INDEX

Pinker, Steven, 219–20
placebo effect, 74
placebos, 74, 176, 179, 182, 184
planchette, 58, 64, 255n10
popular beliefs, 37, 53, 145
post-traumatic stress disorder (PTSD), 12–13
Pratchett, Terry, 174
prayer circle, 84–85
prayers, 34, 36, 41, 250n16, 250n20; of Christian Scientists, 71, 72, 73–74, 78, 80; of disbelief, 42; illnesses and, 71, 73–74; intercessory, 76–77; The Lord's Prayer, 31–32, 37; medical treatment and, 75; power of, 78; as rituals, 37–38; studies on, 76–78
precognition, 20, 161–62, 169, 245n41
Preliminary Assessment, of National Intelligence, 54
Presidential Election, 222
proponents: of dream interpretation, 15; ESP, 166, 167–69; homeopathic, 181, 183; UFO, 54, 55
Proxima Centaurians, 218, 219
pseudoscience, 52, 97, 105–6, 142, 175, 182, 228
psychic energy, 145, 146
psychic fair, 155, 169, 261n25
psychic phenomenon, 161
psychic powers, 117, 155, 164, 173
psychic reading, 156–59, 169–73
psychic scammers, 261n25
Psychic Spies! (CIA-funded project), 163
psychic tricks, 169, 170–71, 172, 173
psychokinesis, 161, 162
psychology, 86, 95; child, 27; of emotions, 211; parapsychology as subfield of, 162, 163, 165, 166, 167, 168, 169; social, 53
Psychology Today (magazine), 13
PTSD. *See* post-traumatic stress disorder
public health, 74, 140, 141
public speaking, 81, 82
pyramidologists, 145

pyramid power, 144–46, 147, 148–49, 150–51, *152*, 153–54, 259n11, 259n12

al-Qaeda hijackers, 217, 222, 232, 266n5
QAnon conspiracy theory, 22, 234
QuackWatch.org (website), 91
Quality-of-Life Index, 132, 135, 257n21, 257n27

Rabin, Yitzhak, 204
Rabinoff, Robert, 131–32, 133–35, 255n7
Radford, Benjamin, 263n2
Radiolab (podcast), 198
Ramsay, Chris, 261n22
Randi, James, 181, 251n12, 253n16
randomization, 124, 125, 138–39, 254n4
rapid eye movements (REM), 11–12, 243n5
razor blades sharpen, 145, 146, 149, 151
Reber, Arthur, 165
recognition, 88–89, 209
reconstruction, of human memory, 5
Reese, Hailey, 65
Reincarnated! (children promoting reincarnation), 163
reincarnation, 17, 62, 128, 163
relationships, 82, 83; abuse in, 86, 136–37, 141; dreams incorporating, 15; exploitative, 86
relaxation response, 140
religious affiliations, 28, 86
religious beliefs, x, 75, 86, 195
religious claims, x, 30–31
religious context, 75
religious conversion experience, 85–86, 94
religious experiences, of children, 31–32
religious groups, 37, 71
religious training, 34
REM. *See* rapid eye movements
replicability, science norms of, 100
responses: fourth, ix–x; relaxation, 140; second, ix
Rhine, Joseph Banks, 162

INDEX

rituals, 27, 28, 29, 89, 90, 93; levitation, 67; pagan, 24; prayer as, 37–38
Roister Doisters (theater club), 68
Rose, Cindy, 145
Roswell, New Mexico, 48, 221
Roswell Army Air Field, 48
Roswell Incident, 48, 221
Rowan, Marcia, 174–77, 179, 181, 182–84, 261n2

Safka, Melanie, 15
Sagan, Carl, 40, 232–33, 240
San Andreas Fault, 111
sanctions, 87
Santa Claus (fictional character), x, 22, 23–24, 27
Savage, Adam, 144
Scarne, John, 168
Scarne's Tricks (Scarne), 168
Schrodt, Philip, 137–38
science, xiii, 184; astrology and, 97–98, 99, 105; biases and, xi; on dreams, 16–17; Hebrew school contrasted with, 40; norms of, 100, 102; pseudo-, 52, 97, 105–6, 142, 175, 182, 228; social, xi, 86, 109, 144; testing, xii; Tyson on, 142–43
Science and Health with Key to the Scriptures (Eddy), 73–74
ScienceBasedMedicine.org (website), 91
science norm of falsifiability, 102
scientific methods, 109, 145, 183, 253n2
scientific skepticism, xi–xii, xiii, 52, 90, 91–92, 160, 240
scientific theory, 90, 97
second responses, ix
second sight, 157
self-appraisals, 99
self-awareness, 13, 17, 18, 20
self-deception, 138, 142, 143
self-fulfilling prophecy, 63, 99
self-identity, 86, 227
semi-sleep, 12–13
sensory information, 12, 20
Sheaffer, Robert, 49

Sheldrake, Rupert, 163
Shermer, Michael, 52, 142, 222–23, 232–33, 240, 264n10
shyness effect, 49
Siddhi meditators, 135, 256n19
single-blinding protocol, 124
SkepticalInquirer.org (website), 91
Skeptic.com (website), 91
The Skeptic Encyclopedia of Pseudoscience (Shermer), 52
skepticism, 19, 65, 123, 144, 183–84, 186; of alternative medicine practices, 177–78; in Christianity, 22; of conspiracy theories, 222–23, 226, 233, 234; constructive, 143; of dowsing, 119–21, 125; about ESP, 160–61, 164, 166–69; about God, 40–41; healthy, xii, 24; scientific, xi–xii, xiii, 52, 90, 91–92, 160, 240
skip codes, 203–4, 208, 266n27
sleep cycle, 12, 244n15
sleep deprivation, 10–11
sleep hallucinations, 19
sleep paralysis, 12–13
sleep phases, 7, 11, 12, 19
sleepwalking, 23
snake oil, 73, 174
Sneeze of Death story, 36–37, 39–40
Snopes.com (website), 91
social consequences, 86
social constructions, 27, 78
social context, 205
social groups, 87–89
social impact theory, 89, 151, 153–54
social interactions, 20, 89, 226
socialization, 27–28, 61–62, 75
social network, 86, 110, 226
social psychology, 53, 99
social science, xi, 86, 109, 144
sociologists, 78, 154, 175, 263n26
Sociology of the Paranormal course, 63, 144, 154
solar systems, 51–52, 218
the soul, 40, 57, 62, 88, 164, 171; the brain and, 17–18; eternality of, 71

INDEX

South Shore Christian Fellowship, 83, 87
spin the bottle (adolescent game), 60
spirit cleansing, 188
spirit medium, 188
spirit orb, 191–92
spirit possession, 19, 65, 67
Stanford University, 109, 111, 113, 114
star constellations, 50, 96, 103–4, 112, 205, 224
storytelling, 35, 62, 103; leveled, 110, 111; sharpened, 110–11
strange noises, 187, 189, 190–91
subjective experience, 111, 149, 179
sun sign astrology, 96, 100, 107
supernatural beliefs, 61, 76, 84, 191, 194, 237
supernatural phenomenon, 91, 151
superpowers, 7, 128, 130
superradiance. *See* Maharishi Effect
Supreme Court, United States, 37

telepathy, 161
telephone game, 110
televangelist, 85
Temple Beth Shalom, 28, 35
tests, for pyramid power, 146–47, 149–54
theater clubs, 68, 83
Thomas, David, E., 266n27
tidal forces, 98, 104–5
time machine, human memory as, 25, 26–27
Transcendental Meditation (TM), 257n21; the brain and, 139, 140, 141; evidence for, 131, 141; Jerusalem study and, *131*, 132, 135–42, 257n25; Maharishi Effect, 132, 133, 134, 135, 137, 138–43, 256n13, 258n40; Maharishi Mahesh Yogi, 129–30, 132, 135; meditation groups, *131*, 132, 135, 136, 139, 256n13; other meditations and, 256n12; Siddhi meditators, 135, 256n19
trauma, 10, 12, 13, 15, 79

true believer, xi, 103, 121
Trump, Donald, 222
Trumpy, Franklin, 134, 135
Twain, Mark, 68
"Two Boats and a Helicopter" (story), 79
Twosday, 198, 201, 205
Tyson, Neil DeGrasse, 43, 142–43

UAP. *See* unidentified aerial phenomenon
UFO. *See* unidentified flying object
Ullman, Dana, 181–83, 184, 262n23, 263n26
UMass Amherst. *See* University of Massachusetts Amherst
uncanny coincidences, x
unconscious hand movements, 64
unidentified aerial phenomenon (UAP), 54, 247n3
unidentified flying object (UFO), xi, 46–47, 48–55, 247n3
Unified Field, 132, 133, 256n13
United States: Air Force, 53; Constitution of, 37; Moon Landings, 221; Supreme Court, 37
unity consciousness, 132–33, 140
University of Iowa, 97, 130, 142, 144
University of Massachusetts Amherst (UMass Amherst), 68, 95, 96, 97, 251n2
unshakeable beliefs, xi

vanity items, 156
Venus, 47, 53
vertigo, 182
vibrations, xii, 156, 193–94
Virgin Mary, 63, 70
vital energy, 178, 184
Vital Vapors (vape shop), 186–88, 189, 190, 191, 194, 195
voodoo, 57–58, 62–63, 142
voodoo dolls (paranormal game), 62–63

water witching, 116
the wave, 89

weather, effected by meditators, 131, 133–34
West, Mick, 49, 50, 222, 233, 247n11
What would Jesus do? (WWJD), 92
whole person treatment, 176
Why People Believe Weird Things (Shermer), 142
Wikipedia (online search engine), 87, 178
willful deception, 92
Williams, Robin, 155
Wiseman, Richard, 167
wisps (translucent waves of light), 7, 187, 190, 192
word-recall task, 163–64
World Trade Center tragedy, 211–12, 216–18, 222, 223, 228–32, 266n5, 267n10

Wuhan, China, 222
WWJD. *See* What would Jesus do?

Yastrzemski, Carl Michael "Yaz," 209–10
youth groups, 88, 90, 92
YouTube (video sharing platform), 53; paranormal-themed, 65; pyramid power on, 146–47; Wiseman on, 167

Zener, Karl, 162
Zener cards, 162
zodiac signs, 17, 99, 100, 104, 253
Zolar's Encyclopedia and Dictionary of Dreams (Fully Revised and Updated for the 21st Century), 16